Phytochemistry of Plants of Genus *Phyllanthus*

Phytochemical Investigations of Medicinal Plants

Series Editor:
Brijesh Kumar

Phytochemistry of Plants of Genus *Phyllanthus*
Brijesh Kumar, Sunil Kumar and K. P. Madhusudanan

Phytochemistry of Plants of Genus *Ocimum*
Brijesh Kumar, Vikas Bajpai, Surabhi Tiwari and Renu Pandey

Phytochemistry of Plants of Genus *Piper*
Brijesh Kumar, Surabhi Tiwari, Vikas Bajpai and Bikarma Singh

Phytochemistry of *Tinospora cordifolia*
Brijesh Kumar, Vikas Bajpai and Nikhil Kumar

Phytochemistry of Plants of Genus *Rauvolfia*
Brijesh Kumar, Sunil Kumar, Vikas Bajpai and K. P. Madhusudanan

Phytochemistry of *Piper betle* Landraces
Vikas Bajpai, Nikhil Kumar and Brijesh Kumar

For more information about this series, please visit: https://www.crcpress.com/Phytochemical-Investigations-of-Medicinal-Plants/book-series/PHYTO

Phytochemistry of Plants of Genus *Phyllanthus*

Brijesh Kumar, Sunil Kumar and
K. P. Madhusudanan

CRC Press is an imprint of the
Taylor & Francis Group, an informa business

First edition published 2020
by CRC Press
6000 Broken Sound Parkway NW, Suite 300, Boca Raton, FL 33487-2742

and by CRC Press
2 Park Square, Milton Park, Abingdon, Oxon, OX14 4RN

CRC Press is an imprint of Taylor & Francis Group, LLC

ISBN: 978-0-367-85755-4 (hbk)
ISBN: 978-0-367-50054-2 (pbk)
ISBN: 978-1-003-01486-7 (ebk)

Typeset in Times
by codeMantra

Contents

Preface

Medicinal plants were used to treat diseases even in prehistoric cultures, having learned about the medicinal properties of plants by trial and error. Ancient civilizations refined these findings leading to the development of traditional medicine systems such as Ayurveda, Unani and Chinese medicine. According to World Health Organization (WHO), more than half of the world's population depends on traditional medicine for their primary healthcare needs. The main reason for this is the high cost of modern medicines and their detrimental side effects. With the increasing demand for herbal medicine, there is an increasing concern about their safety. Availability, efficacy, preservation, quality and standardization are some of the concerns shared by professionals and the general public. Authentication of herbs is often fraught with difficulties. Unscrupulous market practices of adulteration add to the woes. Good agricultural practice (GAP) and good manufacturing practice (GMP) could ascertain the quality of herbal medicines. However, in India, most of the herbal collections are made from their natural habitats, and hence, variables such as climate, geographical location and time of collection influence the quality of herbs. All these parameters have to be factored in when devising means for the quality control of herbs and herbal drugs.

Modern analytical techniques, especially liquid chromatography-mass spectrometry (LC-MS) techniques, play very important roles in the quality control of herbs and herbal drugs. The development of analytical methods involving fingerprinting, quantification, and chemometric analyses is essential for the successful implementation of the quality control processes. Mass spectrometry (MS) coupled with liquid chromatography (LC) has emerged as one of the most favored techniques in recent years for the identification of medicinally active ingredients in complex mixtures of plant extracts. Mass spectrometers having highly sensitive mass analyzers such as time of flight (TOF) provide high-resolution mass data and help in the identification of analytes differing in their exact masses. Quadrupole-ion trap mass spectrometers provide MS^n capability helping in elucidating fragmentation pathways and in quantification of analytes offering very high selectivity and sensitivity. Coupled with the separation power of liquid chromatography, LC-MS allows simultaneous qualitative and quantitative analysis of phytochemicals present at the sub-ppm level in complex matrices of plant extracts.

Genus *Phyllanthus,* an herbal drug in all the traditional systems of medicine, is used primarily as a hepatoprotective in Ayurveda. Its other uses include as a cure for digestive, genitourinary, respiratory, and skin diseases and urolithiasis. *Phyllanthus emblica* or Amla is a dietary source of vitamin C, amino acids and minerals, and is primarily used as an immunostimulator. The major metabolites in these plants are flavonoids, lignans, tannins and other phenols, sterols, terpenoids and hydroxy acids. This book describes the traditional uses, phytochemistry, pharmacological activity and phytochemical analysis of *P. amarus*, *P. niruri*, *P. fraternus* and *P. emblica* and the development of qualitative and quantitative analytical methods for fingerprinting and quantification of the phytoconstituents. Chemometric analyses are described to evaluate the influence of geographical location and the year of collection on the distribution of phytoconstituents and their interspecies variation. Several herbal formulations containing *P. amarus, P. niruri* or, *P. emblica* were also assessed for bioactive components. We are hopeful that this book will be welcomed by researchers working in the field of medicinal plants.

31/12/2019 **The Authors**

Acknowledgments

The completion of this book is due to the Almighty who blessed us with all the resources required to accomplish this journey and the warmest support and helpful advice of many colleagues whom we owe so much. We are glad to express our gratitude to people who have been supportive to us at every step. We express our deep sense of gratitude to Sophisticated Analytical Instrument Facility (SAIF) Division, CSIR-Central Drug Research Institute (CDRI), Lucknow, India, for their supports. All the team members thank the Director of CSIR-CDRI for his support during this period.

Authors

Dr. Brijesh Kumar is a Professor (AcSIR) and Chief Scientist of Sophisticated Analytical Instrument Facility (SAIF) Division, CSIR-Central Drug Research Institute (CDRI), Lucknow, India. Currently, he is facility in charge at SAIF Division of CSIR-CDRI. He has completed his PhD from CSIR-CDRI, Lucknow (Dr. R.M.L Avadh University, Faizabad, UP, India). He has to his credit 7 book chapters, 1 book and 145 papers in reputed international journals. His current area of research includes applications of mass spectrometry (DART MS/Q-TOF LC-MS/ 4000 QTrap LC-MS/ Orbitrap MS^n) for qualitative and quantitative analyses of molecules for quality control and authentication/standardization of Indian medicinal plants/parts and their herbal formulations. He is also involved in the identification of marker compounds using statistical software to check adulteration/substitution.

Dr. Sunil Kumar is currently working as an Assistant Professor at Ma. Kanshiram Government. Degree College, Farrukhabad, Uttar Pradesh, India. He has completed his PhD under the supervision of Dr. Brijesh Kumar on application of mass spectrometric techniques in qualitative and quantitative analyses of phytoconstituents and identification of chemical markers by chemometric technique in *Phyllanthus* and *Rauwolfia* spp. in SAIF at CSIR-CDRI, Lucknow, India. His research interest includes qualitative and quantitative analyses of phytochemicals using LC-MS/MS analysis.

Dr. K. P. Madhusudanan is a Mass Spectrometry Scientist born in 1947 in Kerala, India. He obtained his doctoral degree in 1975 specializing in organic mass spectrometry in National Chemical Laboratory, Pune, India. He worked as a Scientist and Head of SAIF in CSIR-CDRI, Lucknow, India, until 2007. His research experience since 1970 includes various aspects of organic mass spectrometry such as fragmentation mechanism, gas phase unusual reactions, positive and negative ion mass spectrometry of natural products using various ionization techniques, including DART, effects of metal cationization, LC/MS and MS/MS applications and quantitative analysis of drugs and metabolites. He authored more than 150 research publications. He was a member of the editorial board of *Journal of Mass Spectrometry* during 1995–2007. He is a fellow of the National Academy of Sciences, Allahabad, India. At present, he lives in Kochi.

List of Abbreviations and Units

°C	degree celsius
μg	microgram
μL	microliter
APCI	atmospheric pressure chemical ionization
API	atmospheric pressure ionization
BPC	base peak chromatogram
CAD	collision activated dissociation
CE	capillary electrophoresis
CE	collision energy
CID	collision induced dissociation
CXP	cell exit potential
Da	dalton
DAD	diode array detection
DART	direct analysis in real time
DP	declustering potential
EP	entrance potential
ESI	electrospray ionization
FDA	food and drug administration
FIA	flow injection analysis
g	gram
GC-MS	gas chromatography-mass spectrometry
GS1	nebulizer gas
GS2	heater gas
h	hour
HPLC	high-performance liquid chromatography-
ICH	International Conference on Harmonization
IS	internal standard
IT	ion trap
kPa	kilopascal
L	liter
LC	liquid chromatography

LOD	limit of detection
LOQ	limit of quantification
LTQ	linear trap quadrupole
m/z	mass to charge ratio
mg	milligram
min	minute
mL	milliliter
mM	millimolar
MRM	multiple reaction monitoring
MS	mass spectrometry
ms	milli second
MS/MS	tandem mass spectrometry
ng	nanogram
NMPB	national medicinal plants board
NMR	nuclear magnetic resonance
PCA	principal component analysis
PDA	photodiode array detection
psi	pressure per square inch
QqQ_{LIT}	hybrid linear ion trap triple-quadrupole
QTOF	quadrupole time of flight
r^2	correlation coefficient
RDA	retro-Diels-Alder
RSD	relative standard deviation
S/N	signal to noise ratio
SD	standard deviation
t_R	retention time
UHPLC	ultra high-performance liquid chromatography
UV	ultraviolet
WHO	World Health Organization
XIC/EIC	extracted ion chromatogram

1 Introduction *Phyllanthus—Ethno- and Phytopharmacological Review*

1.1 INTRODUCTION

Phyllanthus (*Euphorbiaceae*) is a large genus including more than 800 species (Chaudhary and Rao 2002). It was described by Linnaeus for the first time in 1737 (Webster 1956). "Phyllanthus," derived from Greek meaning "leaf and flower," is named so because the leaf, flower and fruit appear fused. Plants of this genus are trees, shrubs and herbs distributed throughout tropical and subtropical countries of the American, African and Asian continents (Webster 1994). Some of these are also known by different local names such as Stone breaker, Carry me seed, Leaf-flower, Wind breaker, Chanca piedra, Smartweed, Chamberbitter, Gripeweed, Bhoomi amalaki, Dukong anak, Kizharnelli, Meniran, Niruri, Para-parai mi, Phyllanto, Seed-under-leaf, Tamalaka, Yaa tai bai and Zhu zi cao (Natural Medicines 2018).

More than 50 species of *Phyllanthus* including *Phyllanthus amarus*, *P. niruri*, *P. fraternus* and *P. emblica* (synonym: *Emblica officinalis*) are found in India (Tharakan 2011). *P. amarus* is an erect herb growing (30–40 cm tall) throughout India along roads and valleys, on riverbanks, and near lakes. It has tiny greenish leaves, stems and fruits. *P. niruri* is a common weed (30–60 cm tall) found in cultivated fields and wastelands in the rainy season. It is also widespread in coastal areas. *P. fraternus* (Gulf leaf-flower) is an erect annual herb that grows about 45–60 cm tall, and it is a winter weed commonly found in gardens, wastelands and roadsides. *P. emblica* is commonly known as Indian gooseberry (Amla) or emblic leaf-flower which is found throughout India in sunny areas. It is also widely distributed in subtropical and tropical areas of China, India, Indonesia and Malay Peninsula (Zhang et al. 2003; Khan 2009). It is a small to medium-sized deciduous tree. Its fruits have immense medicinal value, and hence they have been consumed as nutritional food and used in traditional medicines for thousands of years in India, China and Southeast Asian countries. *P. emblica* fruits ripen in autumn and are commonly used in the local diet.

1.2 TRADITIONAL USES AND MEDICINAL PROPERTIES OF *PHYLLANTHUS*

Herbs being easily accessible to mankind have been widely explored for their medicinal properties. The cultural tradition associated with the use of medicinal plants for treating various disease conditions is one of the oldest, richest and most diverse in India. This traditional knowledge is codified in the millennia-old Indian system of medicine, Ayurveda. Several *Phyllanthus* species are used widely in traditional medicine for the treatment of diabetes, jaundice, and gall bladder and liver diseases (Calixto et al. 1998; Dhiman and Chawla 2005; Shirani et al. 2017). Plants of these species are known for their remarkable antiviral, antioxidant, antidiabetic, hepatoprotective and anticancer properties, and they have a long history of use in the treatment of intestinal parasites and liver, kidney and bladder diseases (Kirtikar and Basu 1933; Nadkarni and Nadkarni 1976; Kuttan and Harikumar 2011). In traditional medicine, they are used to cure jaundice and other liver diseases. Practitioners of folk and local herbal medicines vouch for the healing powers of this genus for diseases ranging from headache, asthma, diarrhea, skin diseases, diabetes, malaria, indigestion, wounds to a host of others (Mao et al. 2016). It is used in Ayurveda for digestive, genitourinary, respiratory and skin diseases. Several factors such as the therapeutic potential of *Phyllanthus* for the management of many diseases, widespread availability of many species in tropical and subtropical regions, and the diversity of secondary metabolites contribute to their increasing use in traditional medicine. The infusion of leaves, stems and roots of most species of *Phyllanthus* has been used in folk medicine for the treatment of kidney and bladder stones, urinary tract-related diseases, intestinal infections, diabetes and hepatitis B. Seeing the popularity of herbal medicines, the modern society is now eager to resort to green medicines which are without adverse side effects. The genus *Phyllanthus* is one of the most important groups of plants traded as raw herbal drugs in India (Ved and Goraya 2008). Several herbal drug formulations involving *Phyllanthus* species are available in the market in India. For example, Hepex, Liv 52, Livomap, Liv D, Liv Plus, Vimliv, Nirocil, Livocin, Livcure and Livol are very popular herbal drug formulations for jaundice and other liver ailments in general (Figure 1.1).

P. amarus, an important herb of Indian Ayurvedic system, is bitter, astringent, stomachic, diuretic, febrifuge and antiseptic, and used traditionally for its hepatoprotective, expectorant, diuretic, anti-inflammatory, antiseptic and antispasmodial properties. The Indian Ayurvedic system describes its use in curing ailments of stomach, genitourinary system, liver, kidney and spleen. It is also used to treat flu, dropsy, jaundice, diabetes, irregular menstruation, intestinal parasites, asthma, bronchial infections, bladder problems, and hepatic and urolithic diseases (Patel et al. 2011). The other pharmacological properties reported include antipyretic, antibacterial, antioxidant, anticancer and hepatoprotective. It is also reported to inhibit the growth of hepatitis B virus (HBV) and human immunodeficiency virus (HIV) (Ott et al. 1997; Notka et al. 2004). It is predominantly used as a cure for liver disorders in India (Nadkarni and Nadkarni 1976; Kirtikar and Basu 1993; Thyagarajan and Jayaram 1992). Its increasing use can be attributed to its activities against hepatitis B, HIV and other viral diseases, jaundice, kidney stones, gall bladder stones and cancer. It has potent antioxidant properties and thus can scavenge superoxides and hydroxyl radicals (Lim and Murtijaya 2007; Maity et al. 2013).

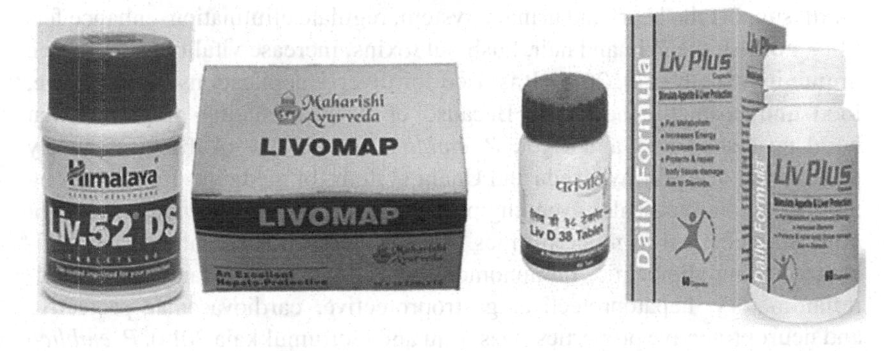

FIGURE 1.1 Some of the herbal products containing *Phyllanthus* available in the Indian market.

P. niruri, bitter in taste, astringent and laxative, is reported to have antidiabetic, antipyretic, antioxidant, carminative, antimicrobial, antiviral, anti-inflammatory, diuretic, antihepatotoxic, antilithic, antihypertensive, anti-HIV and antihepatitis B, antispasmodial, anticancer and antidiarrheal properties. It is also used to treat malaria, jaundice, asthma, edema, ulcers, gonorrhea, syphilis, tuberculosis, cough, diarrhea and vaginitis (Girach et al. 1994; Calixto et al. 1998; Paranjape 2001). It finds use in the treatment of urolithiasis (Barros et al. 2006; Boim et al. 2010). Traditionally, it has been used as a pain reliever, appetite stimulator, antihelminthic and diuretic, and for treating kidney stones, gall bladder stones and liver-related diseases such as liver cancer and jaundice. In Ayurvedic system of medicine, it is used to treat jaundice, urinary diseases, ulcers, skin diseases and diabetes. Ayurvedic preparations of fresh juice and powder are used for the effective treatment of asthma, urinary disorders, bronchitis, anemia, leprosy, cough, blood disorders and jaundice, and for stimulating liver, improving digestion and increasing appetite. The fruits of *P. niruri* can be used for treating tubercular ulcers, wounds and skin infections such as sores, scabies and ring worm (Agharkar 1991).

P. fraternus, an herb possessing astringent, deobstruent, diuretic and antiseptic properties, is used for the treatment of a broad spectrum of diseases including diabetes, alcohol-induced liver damage, jaundice, malaria, kidney stones, genitourinary tract and intestinal infections, hypertension, stroke, abdominal pains, diarrhea and hepatitis (Kirtikar and Basu 1975). It is also used by indigenous people for treating vaginitis and skin eruptions, and is considered analgesic, carminative, digestive, tonic and vermifuge (Taylor 2003). Its antioxidant and anticoagulant properties have also been reported (Koffuor and Amoateng 2011).

The fruits of *P. emblica* are edible, nutritious, adaptogenic and a dietary source of vitamin C, amino acids and minerals. Traditionally, it is used to enhance food absorption, balance stomach acids, fortify liver, nourish the brain, support the heart and urinary system, regulate elimination, enhance fertility, nourish the skin and hair, flush out toxins, increase vitality and enhance immunity (Singh et al. 2011). It is used for diverse applications in healthcare, food and cosmetic industries. Because of its rejuvenating properties that build up lost vitality and vigor, *P. emblica* fruit is one of the most widely used herbal drugs in Ayurveda and Unani systems of medicine. Its very strong antioxidant and radical scavenging properties have made *P. emblica* a sought after nutraceutical in modern times. Recent research has confirmed its antibacterial, hypolipidemic, immunomodulatory, antidiabetic, anticancer, anti-inflammatory, hepatoprotective, gastroprotective, cardiovascular protective and neuroprotective properties (Dasaroju and Gottumukkala 2014). *P. emblica* is an important and versatile ingredient of many traditional medicinal systems such as Chinese herbal medicine, Tibetan medicine and Ayurvedic medicine

(Zhang et al. 2000). Both dried and fresh parts of the plant (fruits, seeds, leaves, root, bark and flowers) are used. Ayurvedic medicinal preparations such as Triphala and Chyawanprash contain *emblica* fruit as a major ingredient. Triphala is used for improving bowel health, digestion, and detoxification and boosting immune system. Chyawanprash is a general tonic that improves mental and physical well-being for people of all ages.

1.3 PHYTOCHEMISTRY

More than 500 compounds comprising alkaloids, flavonoids, glycosides, lignans, phenols, sterols, tannins and terpenoids have been isolated from *Phyllanthus* (Calixto et al. 1998; Qi et al. 2014; Mao et al. 2016; Nisar et al. 2018). Phenolic compounds such as tannins are abundant in *Phyllanthus* species (Nisar et al. 2018). Lignans exist abundantly in the *Phyllanthus* genus. Phyllanthin is an active lignan isolated from various *Phyllanthus* species. The phytochemical diversity in *Phyllanthus* is evident from the classes of compounds identified. Alkaloids, flavonoids, glycosides, lignans, phenols, phenyl propanoids, tannins, terpenes and saponins are the major classes of bioactive compounds found in *Phyllanthus*. The unique structural diversity found among the *Phyllanthus* compounds and their strong bioactive nature make *Phyllanthus* genus of great commercial value.

The phytochemicals in *P. amarus* comprise different classes of organic compounds of medicinal importance. These include alkaloids, flavonoids, hydrolyzable tannins (ellagitannins), major lignans, polyphenols, triterpenes, sterols and volatile oil (Patel et al. 2011; Verma et al. 2014). Active compounds such as phyllanthin, hypophyllanthin, niranthin, nirtetralin, phyltetralin, phyllangin, nirphyllin, phyllnirurin and corilagin have been isolated from *P. amarus*. The leaves contain several lignans such as phyllanthin (a bitter constituent), hypophyllanthin (a non-bitter constituent), niranthin, nirtetralin and phyltetralin (Rastogi and Mehrotra 1991). Isobubbialine and epibubbialine, two new securinega-type alkaloids, were also isolated from the leaves of *P. amarus* along with three known alkaloids, phyllanthine, securinine and norsecurinine (Houghton et al. 1996). From the polar extractives of the aerial parts of *P. amarus*, geraniic acid and other tannins including a novel cyclic hydrolyzable tannin, amarulone, were isolated (Foo and Wong 1992; Foo 1993, 1995). Flavones and sterols were identified in *P. amarus* hexane extract, whereas sterols and triterpenes were identified in its methanol extract (Zubair et al. 2017). An unusual flavonol in natural products, quercetin-3-*O*-*β*-D-glucuronopyranoside, was detected in the aqueous extract of *P. amarus*

TABLE 1.1 Phytochemicals in *P. amarus*

CLASS	*COMPOUNDS*
Alkaloids	Epibubbialine, isobubbialine, norsecurinine, securinine, phyllanthine
Flavonoids	Quercetin, rhamnocitrin, rutin, apigenin, luteolin, kaempferol, astragalin, quercetin-3-*O*-glucoside, quercitrin, 4′,5,7-triethoxy-3,3′,6-trimethoxy flavone
Lignans	Phyllanthin, hypophyllanthin, nirtetralin, phyltetralin, nirtetralin, hinokinin, niranthin, 5-demethoxy niranthin, 7′-oxocubebin dimethyl ether and others
Sterols	Amarosterol A, amarosterol B, cycloeucalenyl acetate, macdougallin, ergosta-5,7,22-trien-3-ol acetate, 17-(1,5-dimethylhexyl)-6-hydroxy-5-methylestr-9-en-3-yl acetate, stigmasterol, *β*-sitosterol, daucosterol, stigmast-5-en-3-ol oleate, 6,7-epoxypregn-4-ene-9,11,18-triol-3,20-dione, 11,18-diacetate, bufalin, 3-acetoxy-7,8-epoxylanostan-11-ol
Tannins and other phenols	Amariin, amariinic acid, amarulone, corilagin, geraniin, geranic acid, furosin, phyllanthusiin D, A, B, C, D, repandusinic acid, elaeocarpusin, 4-galloyl quinic acid, gallic acid
Terpenoids	Phyllanoside, 2Z, 6Z, 10Z, 14E, 18E, 22E-farnesyl farnesol, linalool, phytol, olean-13(18)-ene, methyl ursolate, barringenol R1
Others	Cinnamic acid, 4-hydroxy-3-methoxy-benzoic acid, brevifolin carboxylic acid, rhodopin

(da Fontoura Sprenger and Cass 2013). Table 1.1 lists the major compounds of *P. amarus*.

The phytochemicals similar to those found in *P. amarus* are also present in *P. niruri* and include bioactive molecules such as lignans, flavonoids, glycosides, tannins, alkaloids, ellagitannins, triterpenes, phenylpropanoids and steroids (Bagalkotkar et al. 2006). Seventeen lignans have been identified from *P. niruri*. Phyllanthin, hypophyllanthin, niranthin, phyltetralin, nirtetralin, isonirtetralin, hinokinin, phyllnirurin, lintetralin, isolintetralin, demethylenedioxy niranthin and 5-demethoxy niranthin are among the lignans isolated from *P. niruri*. The number of alkaloids from *P. niruri* is considerably more than that from *P. amarus*. These alkaloids include 4-hydroxysecurinine, 4-methoxydihydronorsecurinine, 4-methoxynorsecurinine, 4-methoxytetrahydrosecurinine, allosecurinine, dihydrosecurinine, *ent*-norsecurinine, epibubbialine, isobubbialine, nirurine, norsecurinine, 4-methoxysecurinine, securinine, securinol A, securinol B and tetrahydrosecurinine. Quercetin-3-*O*-*β*-D-glucopyranosyl-(2→1)-*O*-*β*-D-xylopyranoside, quercitrin, rutin, apigenin-7-*O*-(6″-butyryl-*β*-glucopyranoside), niruriflavone, nirurin, nirurinetin, 7-dimethoxy-3,4′-dihydroxy-3′, 8-di-*C*-prenylflavanone and 6-hydroxy-7,8,2′, 3′, 4′-pentamethoxyisoflavone are some of the flavonoids present in *P. niruri*. Corilagin, geraniin, isocorilagin, phyllanthusiin D, terchebin, prodelphinidin B1, (–)-epicatechin-3-*O*-gallate, (–)-epigallocatechin 3-*O*-gallate and 1-*O*-galloyl-6-*O*-luteoyl-*α*-D-glucose are some of the tannins found. 2,3,5,6-Tetrahydroxybenzyl acetate and phyllangin were also isolated from *P. niruri* (Wei et al. 2004). Other constituents present include sterols, terpenoids, phenols and organic acids (Calixto et al. 1998; Narendra et al. 2012). Six phenolic compounds, namely, gallic acid, (–)-epicatechin, (+)-gallocatechin, (–)-epigallocatechin, (–)-epicatechin-3-*O*-gallate and (–)-epigallocatechin 3-*O*-gallate, were isolated from hairy roots of *P. niruri* (Ishimaru et al. 1992). The major phytoconstituents of *P. niruri* are listed in Table 1.2.

The plant extract of *P. fraternus* is reported to contain alkaloids, terpenoids, steroids, alkamides, seco-sterols and lignans (Sittie et al. 1998; Gupta and Ali 1999; Habib-ur-Rehman et al. 2007). However, the number of components identified is much less compared to *P. amarus* and *P. niruri*. Five securinega alkaloids, namely, (+)-allonorsecurinine, *ent*-norsecurinine, nirurine, bubbialine and epibubbialine, and a lignan phyllanthin were isolated from *P. fraternus* (Komlaga et al. 2017). Tannins and flavonoids are not reported. From the roots of *P. fraternus*, four new seco-sterols, phyllanthosterol, phyllanthosecosteryl ester, phyllanthostigmasterol and fraternusterol, were isolated (Gupta and Ali 1999). Two new oxygenated heterocyclic compounds, phyllanthusolactone and phyllanthodocosanyl ester, were also isolated from *P. fraternus* roots (Gupta and Ali 2000). From the methanol extract of the aerial parts of *P. fraternus*, a new fatty acid and a new acyl tetraglucoside along with known aliphatic alcohol, palmityl glucuronoside and steroids were isolated (Ali et al. 2016). The main phytochemical constituents of *P. fraternus* are listed in Table 1.3.

Several reports review the phytochemicals in *P. emblica* (Singh et al. 2011; Patel and Goyal 2012). Alkaloids, carbohydrates, essential oils, fatty acids, flavonoids, glycosides, lignans, phenols, tannins, sterols, ascorbic acid, music acid and terpenoids are the phytochemical constituents of *P. emblica*. It is reported to contain phenolic constituents such as gallic acid, corilagin and chebulagic acid (Zhang et al. 2001a, 2004; Kumaran and Joel Karunakaran 2006); hydrolyzable tannins like emblicanin A and B (Ghosal 1996); flavonoids like quercetin (Anila and Vijayalakshmi 2002); and alkaloids like phylantine and phylantidine (Khanna and Bansal 1975). From the aqueous acetone extract of *P. emblica* fruit juice powder, six new gallates, L-malic acid 2-*O*-, mucic acid 2-*O*-, mucic acid 1,4-lactone 2-*O*-, 5-*O*-, 3-*O*- and 3,5-di-*O*-gallates, were isolated (Zhang et al. 2001a). In another study, six new ellagitannins, phyllanemblinins A–F, and 30 known tannins and related compounds were isolated

TABLE 1.2 Phytochemicals in *P. niruri*

CLASS	*COMPOUNDS*
Alkaloids	4-Hydroxysecurinine, 4-methoxydihydronorsecurinine, 4-methoxynorsecurinine, 4-methoxytetrahydrosecurinine allosecurinine, dihydrosecurinine, ent-norsecurinine epibubbialine, Isobubbialine, nirurine, norsecurinine, phyllanthine, securinine, securinol A, securinol B, tetrahydrosecurinine, 1,12-diazacyclodocosane-2,11-dione
Flavonoids	Quercetin-3-*O*-*β*-D-glucopyranosyl-(2→1)-*O*-*β*-D-xylopyranoside, quercitrin, rutin, apigenin-7-*O*-(6″-butyryl-*β*-glucopyranoside), niruriflavone, nirurin, nirurinetin, 7-dimethoxy-3,4′-dihydroxy-3′,8-di-C-prenylflavanone, 6-hydroxy-7,8,2′,3′, 4′-pentamethoxyisoflavone
Lignans	Phyllanthin, hypophyllanthin, linnanthin, niranthin, nirphyllin, isolintetralin, lintetralin, neonirtetralin, nirtetralin, nirtetralin A, nirtetralin B, phyltetralin, seco-4-hydroxy-lintetralin, urinatetralin, demethylenedioxyniranthin, dihydrocubebin, hydroxyniranthin, cubebin dimethyl ether, hinokinin, phyllnirurin and others
Sterols	24-Isopropylcholesterol, *β*-sitosterol
Tannins and other phenols	1,2,4,6-Tetra-*O*-galloyl-*β*-D-glucose,1,2-Di-*O*-galloyl-3,6-(R)-hexa-hydroxydiphenoyl-*β*-D-glucose, corilagin, geraniin, isocorilagin, elaeocarpusin, phyllanthusiin D, terchebin, prodelphinidin B1, (–)-epicatechin-3-*O*-gallate, (–)-epigallocatechin-3-O-gallate, 1-*O*-galloyl-6-*O*-luteoyl-*α*-D-glucose, ellagic acid, gallic acid, catechin, epicatechin, gallocatechin, elpigallocatechin
Terpenoids	Phyllanthenol, phyllanthenone, phyllantheol, friedelin, 3,7,11,15,19,23-hexamethyl-2Z,6Z,10Z,14E,18E,22E-tetracosahexen-1-ol, orthosiphol G, orthosiphol I, trans-phytol
Others	4-*O*-Caffeoylquinic acid, brevifolin carboxylic acid, ethyl brevifolin carboxylate, methyl brevifolin carboxylate, phyllangin, 2,3,5,6-tetrahydroxybenzyl acetate

(Zhang et al. 2001b). Two new flavonoids, kaempferol-3-*O*-*α*-L-(6″-methyl)-rhamnopyranoside and kaempferol-3-*O*-*α*-L-(6″-ethyl)-rhamnopyranoside, were also isolated from *P. emblica* (Habib-ur-Rehman et al. 2007). Sixteen compounds including sterols, phenols, tannins and pyrano [3, 2-b] pyran-2, 6-dione were isolated from the 95% ethanol extract of the bark of wild *P. emblica* (Yang et al. 2014). Using silica gel, polyamide and Sephadex LH-20 chromatography, 13 compounds including triacontanol, triacontanoic acid, *β*-amyrin ketone, *β*-amyrin-3-palmitate, betulonic acid, betulinic acid, ursolic

TABLE 1.3 Phytochemicals in *P. fraternus*

CLASS	*COMPOUNDS*
Alkaloids	(+)-allonorsecurinine, ent-norsecurinine, bubbialine epibubbialine, nirurine, E,E-2,4-octadienamide, E,Z-2,4-decadienamide
Lignans	Phyllanthin
Sterols	Fraternusterol, phyllanthosecosteryl ester, phyllanthosterol phyllanthostigmasterol, *β*-sitosterol, *β*-sitosteryl oleate, *β*-sitosteryl linoleate, stigmasterol.
Terpenoids	Phyllanterpenyl ester, olean-18-en-3*α*-ol
Others	1-Triacontanaol, 1-pentacosanol, cis-*n*-octacos-17-enoic acid, n-dodecanoyl-*O*-*β*-D-glucopyranosyl-(2′→1″)-*O*-*β*-D-glucopyranosyl-(2″→1‴)-*O*-*β*-D-glucopyranosyl-(2‴→1⁗)-*O*-*β*-D-glucopyranoside, palmityl glucuronoside, phyllanthusolactone, phyllanthodocosanyl ester, *n*-octacosenoic acid, lauryl tetraglucoside.

acid and oleanolic acid were isolated and characterized. Compared to *P. amarus* and *P. niruri*, alkaloids and lignans are much less in *P. emblica*, whereas a large number of flavonoids, tannins, phenols/polyphenols, terpenoids and organic acid derivatives are present in *P. emblica*. This justifies *P. emblica*'s prime use as a food supplement and immunostimulator in traditional medicine. Table 1.4 lists the major components in *P. emblica*.

TABLE 1.4 Phytochemicals in *P. emblica*

CLASS	*COMPOUNDS*
Alkaloids	Methyl (2S)-1-[2-(furan-2-yl)-2-oxoethyl]-5-oxopyrrolidine-2-carboxylate, 5-hydroxy-isoquinoline
Flavonoids	Avicularin, isoquercitrin, kaempferol, kaempferol-3-*O*-*α*-L-(6″-ethyl)-rhamnopyranoside, kaempferol-3-*O*-*α*-L-(6″-methyl)-rhamnopyranoside, kaempferol-3-*O*-*β*-D-gluco-pyranoside, quercetin, quercetin-3-*O*-*β*-D-glucopyranoside, apigenin-7-*O*-(6″-butyryl-*β*-glucopyranoside), (S)-eriodictyol 7-*O*-(6″-*O*-(E)-*β*-oumaroyl)-*β*-D-glucopyranoside, (S)-eriodictyol 7-*O*-(6″-*O*-galloyl)-*β*-D-glucopyranoside, catechin, (–)-epiafzelechin, (–)-epicatechin, (–)-epigallocatechin, (+)-gallocatechin, emblirol A, emblirol B
Lignans	Isolariciresinol, 4-ketopinoresinol, lirioresinol A, medioresinol, syringaresinol, 4,9,9′-trihydroxy-3,4′-dimethoxy-8-*O*-3′-neolignan
Sterols	Stigmast-4-en-3-one, stigmast-4-en-3,6-dione, stigmast-4-en-6*β*-ol-3-one, stigmast-4-ene-3*β*,6*α*-diol, *β*-daucosterol, stigmasterol, *β*-sitosterol

(*Continued*)

TABLE 1.4 (*Continued*) Phytochemicals in *P. emblica*

CLASS	*COMPOUNDS*
Tannins and other phenols	Carpinusin, chebulagic acid, chebulanin, corilagin, furosin, geraniin, isocorilagin, isomallotusinin, isostrictinin, mallonin, mallotusinin, neochebulagic acid, phyllanemblinin A, phyllanemblinin B, phyllanemblinin C, phyllanemblinin D, phyllanemblinin E, phyllanemblinin F, phyllanthunin, punicafolin, putranjivain A, putranjivain B, tercatain, epicatechin-(4*β*→8)-epigallocatechin, phyllemtannin, prodelphinidin B1, prodelphinidin B2, prodelphinidin B-2,3′-*O*-gallate, 1,2,4,6-tetra-*O*-galloyl-*β*-D-glucose, (–)-epigallocatechin 3-*O*-gallate, 1-*O*-galloyl-*β*-D-glucose, 4-*O*-methylellagic acid-3′-*α*-rhamnoside, chebulic acid, ellagic acid, methyl gallate, ethyl gallate, 1-ethyl 6-methyl ester 2-*O*-gallate, flavogallonic acid bislactone, gallic acid, gallic acid-3-*O*-(6′-*O*-galloyl)-*β*-D-glucoside, gallic acid-3-*O*-*β*-D-glucoside, 1,6-di-*O*-galloyl-*β*-D-glucose, mucic acid 1,4-lactone 2-*O*-gallate, mucic acid 1,4-lactone 3,5-di-*O*-gallate, mucic acid 1,4-lactone 3-*O*-gallate, mucic acid 1,4-lactone 5-*O*-gallate, mucic acid 1,4-lactone methyl ester 5-*O*-gallate, mucic acid 1,4-lactone 6-methyl ester 2-*O*-gallate, mucic acid 1,4-lactone 6-methyl ester 5-*O*-gallate, mucic acid 1-methyl ester 2-*O*-gallate, mucic acid 2-*O*-gallate, mucic acid-3-*O*-gallate, mucic acid 6-methyl ester 2-*O*-gallate, L-malic acid 2-*O*-gallate, mucic acid di-methyl ester 2-*O*-gallate, 3,4,8,9,10-pentahydroxy-dibenzo[b,d]-3 pyran-6-one, 3-ethylgallic acid, ethyl gallate, methyl gallate, 3, 4, 3′-*O*-trimethylellagic acid, lup-20(29)-ene-3β, 16β-diol, 3-*O*-methylellagic acid-4′-*O*-*α*-L-rhamnoside, gallic acid, gallocatechin, epigallocatechin, 3, 3′-*O*-dimethylellagic acid-4′-*O*-*α*-L-rhamnoside and 3, 3′, 4-*O*-trimethylellagic acid-4′-*O*-*β*-D-glucoside
Terpenoids	3,20-Dioxo-dinorfriedelane, lupeol, *β*-Amyrin ketone, 4′-hydroxyphyllaemblicin B, Glochicoccin D, phyllaemblic acid, phyllaemblic acid B, phyllaemblic acid C, phyllaemblicin A, phyllaemblicin B, phyllaemblicin C, phyllaemblicin D, phyllaemblicin E, phyllaemblicin F, phyllaemblicin G1, phyllaemblicin G2, phyllaemblicin G3, phyllaemblicin G4, phyllaemblicin G5, phyllaemblicin G6, phyllaemblicin G7, phyllaemblicin G8, phyllaemblinol, *β*-caryophyllene, betulonic acid, betulinic acid, betulin
Others	Cinnamic acid, coniferyl aldehyde, 2-(2-methylbutyryl) phloroglucinol, 1-*O*-(6″-*O*-*β*-D-apiofuranosyl)-*β*-D-glucopyranoside, 2,6-dimethoxy-4-(2-hydroxyethyl)phenol 1-*O*-*β*-D-glucopyranoside, 2-carboxylmethylphenol 1-*O*-*β*-D-glucopyranoside, 4-hydroxy-3-methoxybenzaldehyde, methyl caffeate, methyl-4-hydroxybenzoate, pyrogallol, syringaldehyde, vanillic acid, triacontanol, triacontanoic acid, pyrano [3, 2-b] pyran-2, 6-dione, 5-hydroxymethylfurfural

1.4 PHARMACOLOGICAL ACTIVITY

The medicinal plants under the genus *Phyllanthus* produce diverse classes of secondary metabolites including alkaloids, flavonoids, lignans, phenolic acids and tannins, which are used for treating a variety of medical conditions (Calixto et al. 1998). Pharmacological screening revealed that phyllanthin has antioxidant, hepatoprotective, anticancer, antidiabetic, immuno-suppressant and anti-inflammatory properties. Phyllanthin and hypophyllanthin show antitumor activities. It is reported that the aqueous extracts of several species of *Phyllanthus* exhibit potent *in vitro* and *in vivo* inhibitions against HBV (Syamasundar et al. 1985; Shead et al. 1992; Ott et al. 1997). The effectiveness and bio-safety of the genus *Phyllanthus* for chronic HBV infection was reported (Liu et al. 2001). It is an effective remedy for urolithiasis, one of the oldest known diseases (López and Hoppe 2010). The presence of a large number of polyphenols in *Phyllanthus* species is the reason why they are strong antioxidants (Joy and Kuttan 1995).

A wide spectrum of pharmacological activities are exhibited by *P. amarus* such as antibacterial, anticancer, antidiarrheal, gastroprotective, antiulcer, antifungal, analgesic, anti-inflammatory, antiallodynic and antioedematogenic, antinociceptic, antioxidant, antiplasmodial, antiviral, aphrodisiac, contraceptive, diuretic and antihypertensive, hepatoprotective, hypoglycemic, hypocholesterolemic, immunomodulatory, nephroprotective, radioprotective and spasmolytic (Patel et al. 2011). It was reported that the aqueous extract of *P. amarus* was active against 20-methylcholanthrene (20-MC)-induced sarcoma development (Rajeshkumar et al. 2002). Administration of the methanolic extract of *P. amarus* significantly reduced the toxic side effects of cyclophosphamide in mice without interfering with its antitumor efficiency (Kumar and Kuttan 2005). Lignans have shown antimitotic, antitumor, antiviral (MacRae and Towers 1984) and anti-HIV Schroder et al. 1990) activities, and protective effects against hormone-related cancers due to their antioxidant effects (Lee and Xiao 2003). The methanol extract of *P. amarus* was found to have very high inhibitory activity against bacterial species such as *Pseudomonas aeruginosa*, *Klebsiella pneumoniae*, *Proteus mirabilis*, *Streptococcus faecalis*, *Enterobacter* species, *Serratia marcescens*, *Staphylococcus aureus* and *Escherichia coli*. The aqueous extract of *P. amarus* and its ethanol fraction showed hypotensive effect in rabbits (Amonkan et al. 2013). Inhibitory action was also noticed for aqueous and methanol extracts against *Leptospira*. The methanol extract of the leaves of *P. amarus* has great anti-inflammatory and analgesic potential (Ofuegbe et al. 2014). *P. amarus*, widely used in Ayurvedic medicine for the treatment of liver diseases, has shown both *in vitro* and *in vivo* activities against HBV. A correlation between total phenolic content and antioxidant activity was observed for the methanolic extract of *P. amarus* (Kumaran and Joel Karunakaran 2007). Hepatoprotective effect of the aqueous extract and antidiabetic effect of the ethanolic extract have been reported (Pramyothin et al. 2007). Ellagitannins and flavonoids from *P. amarus* relieved the oxidative stress after radiation (Londhe et al. 2009). Ellagitannins geraniin and amariin were found to protect liver cells from ethanolic cytotoxicity (Londhe et al. 2012). The aqueous ethanol extract of *P. amarus* was found to block HIV-1 attachment both *in vitro* and *in vivo* (Notka et al. 2004). The lignan niranthin exhibited anti-inflammatory and antiallodynic activities (Kassuya et al. 2006). It was suggested that the polyphenols and lignans in *P. amarus* could be the major contributors to the immunomodulatory effect of the plant (Jantan et al. 2014). Methanolic extract of *P. amarus* showed significant concentration-dependent antibacterial activity particularly against gram-negative bacteria (Mazumder et al. 2006). Silver nanoparticles synthesized from the *P. amarus* extract exhibited excellent antibacterial potential against multidrug-resistant strains of *P. aeruginosa* from burn patients (Singh et al. 2014). Extensive studies on *P. amarus* have demonstrated it to be antiviral against hepatitis B and C viruses, hepatoprotective, immunomodulating and anti-inflammatory (Thyagarajan et al. 2002).

A recent review summarizes the scientific evidence of the pharmacological properties of *P. niruri* (Lee et al. 2016). The richness of efficient medicinal metabolites such as alkaloids, flavonoids, lignans, polyphenols, tannins and triterpenes renders *P. niruri* with therapeutic activities including antiviral, antimicrobial, antihepatic, antitumor, antioxidant and antidiabetic activities (Bagalkotkar et al. 2006). One of the active components of *P. niruri* is niruriside found responsible for antiviral activity against HIV. Geraniin and corilagin were also found to be active against HIV. The alkaloids in *P. niruri* were found to have antispasmodic activity. It also stops the progression of non-alcoholic fatty liver diseases and atherosclerosis (Al Zarzour et al. 2017). The lignans phyllanthin and hypophyllanthin exhibited antitumor activities. The preventive effect of the aqueous extract of *P. niruri* in stone formation (nephrolithiasis) and easing its elimination is demonstrated (Boim et al. 2010). *P. niruri* intake contributed to the elimination of urinary calculi (Pucci et al. 2018). Water and methanol extracts of *P. niruri* and *P. amarus* were found to have significant antibacterial and antioxidant activities (Poh-Hwa et al. 2011). Aqueous and methanolic extracts of *P. niruri* exhibited hepatoprotective and antioxidant activities (Harish and Shivanandappa 2006; Amin et al. 2013). Nirtetralin and niranthin were tested against human HBV *in vitro* and found to suppress effectively expression of both HBsAg and HBeAg

(Huang et al. 2003; Wei et al. 2012). The ethanolic, dichloromethane and aqueous extracts of *P. niruri* exhibited *in vitro* and *in vivo* antiplasmodial activities (Tona et al. 1999, 2001). Lipid-lowering effect of *P. niruri* is demonstrated in hyperlipidemic rats (Khanna et al. 2002). The nematocidal effect exhibited by the hexane extract of *P. niruri* was found to be due to prenylated flavanones (Shakil et al. 2008). It inhibits calcium oxalate crystallization and thus interferes with growth and aggregation of calcium oxalate crystals in urine (Freitas et al. 2002; Boim et al. 2010).

The antimalarial activity of *P. fraternus* appears to be due to the presence of securinega alkaloids (Komlaga et al. 2017). It was demonstrated that the extract of *P. fraternus* had significant antiplasmodial activity against the *Plasmodium* parasites (Komlaga et al. 2016). The two isomeric alkamides isolated from this plant were found to have moderate antiplasmodial activity in an *in vitro* assay (Sittie et al. 1998). *In vivo* testing of the antimalarial activity of the *P. fraternus* extract against *Plasmodium berghei* showed positive results for suppression of parasitamea (Matur et al. 2009). The ethanolic extract of *P. fraternus* exhibited antioxidant and anticoagulant effects (Koffuor and Amoateng 2011; Upadhyay et al. 2014). This supports its traditional uses in the management of conditions where oxidative stress is implicated in thromboembolic stroke (Koffuor and Amoateng 2011). Considerable antioxidant and antimicrobial properties were found in the hexane and ethyl acetate fractions of *P. fraternus* (Ibrahim et al. 2017). The aqueous extract of *P. fraternus* exhibited ameliorative effects on vascular complications caused by cyclophosphamide in rats (Moirangthem et al. 2016). Hepatoprotective effect on lead-induced toxicity has been reported (Shah and Jain 2016). Antiviral activity of the *P. fraternus* extract against hepatitis virus has been demonstrated (Calixto et al. 1998). Aqueous and ethanolic extracts of *P. fraternus* exhibited significant anti-inflammatory activities (Oseni et al. 2013).

A recent report concludes that the fruits of *P. emblica* are a potential source of natural antioxidants useful for hepato-, cyto- and radioprotection as well as reducing oxidative stress in many pathological conditions (Charoenteeraboon et al. 2010). It is antioxidant (Dasaroju and Gottumukkala 2014), hypolipidemic (Santoshkumar et al. 2013), hypoglycemic (D'souza et al. 2014), antimicrobial (Rani and Khullar 2004), anticancer (Liu et al. 2012) and anti-inflammatory (Golechha et al. 2014; Wang et al. 2017). In a recent review, the therapeutic applications of *P. emblica* are discussed in detail (Bhandari and Kamdod 2012). Other reviews discuss the phytochemistry, pharmacology and medicinal properties of *P. emblica* (Patel et al. 2011; Gaire and Subedi 2014). The ethyl acetate fraction of the methanol extract of *P. emblica* showed significant antioxidant, immunomodulatory and anticancer activities (Liu et al. 2008; Luo et al. 2009). The butanol extract of the water-soluble fraction of the fruits of *P. emblica* was found to have cytoprotective effect on gastric ulcer formation (Bandyopadhyay et al. 2000). A brief overview is reported listing the evidence supporting anticancer activity of Indian gooseberry extracts (Zhao et al. 2015). Hepatoprotective activity of *P. emblica* is reviewed (Thilakchand et al. 2013). Sesquiterpenoid glycosides from *P. emblica* exhibited potential anti-HBV activities (Lv et al. 2014).

1.5 PHYTOCHEMICAL ANALYSIS

Herbal raw materials are influenced by factors such as identification of the plant, geographic location, seasonal variation, and drying and storage conditions. There is an increasing concern regarding adulteration and species admixture in the raw herb trade as the adverse consequences on the health and safety of consumers cannot be ignored (Srirama et al. 2017). In order to make herbal medicine more credible and acceptable, a comprehensive herbal product authentication has to be evolved. Qualitative and quantitative analysis of a number of marker compounds is the basis of quality control of herbal medicines. This approach fails when there are no suitable markers. Moreover, selecting a few compounds from a complex mixture may not reflect the real picture as synergistic effects are not taken into account. Herbal drugs contain diverse compounds in complex matrices, and the overall therapeutic efficacy may not depend on a single compound but on the synergy of several components. Chemical fingerprinting or evaluation of a product in its entirety can be effectively used to describe the complexity of herbal medicines (Liang et al. 2004). It serves as a tool for identification, authentication and quality control of herbal drugs all over the world.

Analysis of such complex data requires special software, and Principal Component Analysis (PCA) is one such software used for chemometric analysis, an approach to the interpretation of patterns in multivariate data, to evaluate similarities and to discriminate herbs from closely related species and adulterants (Martins et al. 2011; Goodarzi et al. 2013; Gad et al. 2013). Data of TLC, HPLC, GC, GC-MS and HPLC-MS analyses can also be used to generate a chemical fingerprint. Because of its simplicity and reliability, chromatographic and spectral fingerprinting technique is a popular and potent tool for quality control of herbal medicines (Balammal et al. 2012; Joshi 2012).

Most of the active components in *Phyllanthus* belong to phenolics and lignans. Until recently, their analysis had been carried out using HPTLC, HPLC, GC and GC-MS. But lately, because of the extremely high sensitivity and specificity, LC-MS or LC-MS/MS has found its place in the analysis of phytoconstituents of *Phyllanthus*. There are several reports of use of HPTLC for the analysis of *Phyllanthus* constituents, especially for lignans (Annamalai and Lakshmi 2009; Tripathi et al. 2006). HPTLC analysis was used by several workers for the estimation of lignans and other bioactive components (Nayak et al. 2010; Mehta et al. 2013), alkaloids, flavonoids, glycosides and saponins (Shah et al. 2017) in *Phyllanthus*. A TLC image analysis method was developed for detecting and quantifying bioactive phyllanthin in *P. amarus* and commercial herbal drugs (Ketmongkhonsit et al. 2015). Comparative quantification of phyllanthin in different parts of *P. amarus* plants was achieved by HPTLC and HPLC (Annamalai and Lakshmi 2009). Simultaneous determination of four lignans from *Phyllanthus* was carried out by an HPTLC method using chiral TLC plates (Srivastava et al. 2008). Phytochemical diversity of *P. amarus* was assessed by HPTLC fingerprints (Khan et al. 2011). Twelve lignans originating from *Phyllanthus* were separated by micellar electrokinetic chromatography (Kuo et al. 2003).

Several bioactive components were identified in *P. amarus* extracts by GC-MS (Arun et al. 2012; Mamza et al. 2012). Hexane and methanol extracts of the leaves of *P. amarus* were analyzed by GC-MS (Zubair et al. 2017). Nine compounds, mainly flavones and sterols, were identified from the hexane fraction, and 14 compounds including steroids and triterpenoids were identified in the methanol extract. More than 50 phyto-compounds were detected from the methanolic extract of *P. fraternus* by GC-MS (Singh 2016). Terpenes, phytosterols and terpenoids were identified in the ethyl acetate extract of *P. emblica* by GC-MS (Deepak and Gopal 2014; El Amir et al. 2014). GC-MS analysis of the chloroform extract of the leaves of *P. amarus* identified a few fatty acids, esters and hydrocarbons (Igwe and Okwunodulu 2014). H-NMR-based metabolomics approach using partial least square (PLS) results showed that phytochemicals, including hypophyllanthin, catechin, epicatechin, rutin and quercetin, and chlorogenic, caffeic, malic and gallic acids were correlated with antioxidant and α-glucosidase inhibitory activities of the *P. niruri* extract (Mediani et al. 2017). Several HPLC methods are reported for the estimation of phyllanthin in *P. amarus* (Hamrapurkar and Pawar 2009). A HPLC method was developed for the quantification of phyllanthin and hypophyllanthin in *P. amarus* (Murali et al. 2001). A validated HPLC method was reported for the estimation of phyllanthin in the *P. amarus* extract (Alvari et al. 2011). The leaves of *P. amarus* were found to contain the highest amounts of phyllanthin (0.7% w/w) and hypophyllanthin (0.3% w/w) as compared to the other parts of the plant based on a RP-HPLC method (Sharma et al. 1993). Phyllanthin and hypophyllanthin in herbal products containing the *P. niruri* extract were determined by HPLC-UV detection (Rai et al. 2009). Authentication of *P. niruri* from related species was done using HPLC fingerprint and simultaneous quantitative analysis of phyllanthin and hypophyllanthin (Nasrulloh et al. 2018). Using corilagin as a marker, a validated HPLC method was reported for the standardization of *P. niruri* herb and commercial extracts in Brazil (Colombo et al. 2009). A sensitive HPLC-fluorescence detection method was reported for the simultaneous determination of four lignans, phyllanthin, hypophyllanthin, phyltetralin and niranthin, from *P. niruri* in rat plasma (Murugaiyah and Chan 2007). Phenolic constituents in the aqueous ethanolic extracts of *P. niruri* and *P. urinaria* were identified by HPLC assay (Mahdi et al. 2011). A HPLC method was developed for the simultaneous estimation of ascorbic and gallic acids in the *P. emblica* fruit extract prepared from freeze-dried juice (Sawant et al. 2010). In connection with a study on the antioxidant activity of *P. emblica* fruit juice, six polyphenol compounds were identified by HPLC analysis (Liu et al. 2008). Lignans such as phyllanthin, hypophyllanthin, niranthin and the antioxidant ellagic acid in the methanolic extracts of *P. amarus* were quantified by LC/MS (Muthusamy et al. 2018). The simultaneous determination of six lignans of therapeutic importance (heliobuphthalmin lactone, virgatusin, hypophyllanthin, phyllanthin, nirtetralin and niranthin) in four *Phyllanthus* spp. using HPLC-PDA-MS was reported (Shanker et al. 2011). LC-MS/MS analysis of the ethyl acetate fraction of the 70% ethanolic extract of *P. amarus* led to the identification of 28 different phenolic compounds (Maity et al. 2013). Phytochemical analysis of *P. amarus* using HPLC-UV-MS and LC-MS led to the identification of a number of phenolic acids, flavonoids and phytosterols (Corciovă et al. 2018). A comparison of four *Phyllanthus* species including *P. amarus* and *P. niruri* was reported by characterizing the chemical profiles of their aqueous extracts using LC-ITMS^n (da Fontoura Sprenger and Cass 2013). Twenty phenolic compounds were identified in enriched phenolic extracts of *P. acuminatus* by UPLC-ESI-MS (Navarro et al. 2017), whereas eight phenolic compounds were simultaneously quantified in *P. simplex* using HPLC-DAD-ESI-MS (Niu et al. 2012) and four lignans from *P. urinaria* were simultaneously determined in rat plasma (Fan et al. 2015). During HPLC-MS/MS analysis of *P. emblica* extract and fractions, 144 peaks were detected, of which 67 were tentatively identified mostly as ellagitannins, flavonoids and simple gallic acid derivatives (Yang et al. 2012). Tannins, organic acids and flavonoids of *P. emblica* extract could be identified by HPLC-ESI-MS/MS method during a run time of 120 mins (Wang et al. 2017). LC-MS analysis showed that gallic acid, quinic acid, quercetin and other flavonoids were the major constituents in *P. emblica* extracts (Packirisamy et al. 2018). Geraniin, quercetin-3-β-D-glucopyranoside, kaempferol-3-β-D-glucopyranoside, isocorilagin, quercetin and kaempferol were identified by spectral methods (Liu et al. 2008).

Recent reports from our laboratory describe the effective use of high-performance liquid chromatography coupled with quadrupole time of flight- mass spectrometry (HPLC-ESI-QTOF-MS/MS) for the identification and characterization of 52 phenolics and lignans in the ethanolic extract of *P. amarus* (Kumar et al. 2015a). Six bioactive compounds were quantified using UPLC-ESI-MS/MS in a run time of 3.6 mins with the limit of quantitation (LOQ) of 0.44–3.82 ng/mL. HPLC-ESI-QTOF-MS/MS was used for the identification and characterization of phenolics and terpenoids from ethanolic extracts of four *Phyllanthus* species tentatively identifying 30 compounds and unambiguously identifying 21 compounds by comparison with their authentic standards (Kumar et al. 2017a). Using UPLC-ESI-MS/MS, 23 targeted bioactive compounds in ethanolic extracts of *P. amarus*, *P. niruri*, *P. fraternus* and *P. emblica* and their herbal products were quantified simultaneously in polarity-switching MRM mode (Kumar et al. 2017b). Geographical variation of phytoconstituents in *P. amarus* was investigated by a rapid and versatile direct analysis in real time-time of flight-mass spectrometry (DART-TOF-MS) method (Kumar et al. 2017c). Sixteen constituents including alkaloids and lignans were identified, and the ten chemical markers identified by PCA could discriminate between the samples of different geographical locations. These studies form part of our attempt to chemical fingerprinting (identify, characterize and quantify the phytoconstituents) of *P. amarus*, *P. niruri*, *P. fraternus* and *P. emblica*. Along with the species identity, geographical variations influence the efficacy of herbal drugs. Development of discriminative analytical methods for authentication, phytochemical fingerprinting and geographical distribution is absolutely necessary for the quality control of raw and processed herbal products.

Screening of Phytochemicals in *P. amarus* by HPLC-ESI-QTOF-MS/MS

2

2.1 METHODS USED FOR ANALYSIS

Generally, the aerial parts of *Phyllanthus amarus* are utilized in the preparations of various herbal formulations. Variations in soil conditions, climate and geographical region play crucial roles in the contents of secondary metabolites. This will, in turn, influence the efficacy of herbal drugs. Discriminative analytical methods are needed for authentication, phytochemical fingerprinting and assessment of the geographical variations of phyto-components in both raw and processed *P. amarus* to ensure their quality and efficacy.

The major constituents in *P. amarus* belong to alkaloids, flavonoids, lignans and polyphenols. Liquid chromatography-tandem mass spectrometry (LC-MS/MS) should be the technique of choice to analyze *P. amarus* metabolites both qualitatively and quantitatively. LC-QTOF-MS gives accurate mass and MS/MS data, and thus helps identify and characterize the components, whereas UPLC-QTRAP-MS provides fast and robust quantitative data to assess the contents of the various phytoconstituents. UPLC-QTRAP-MS/MS in MRM acquisition mode has drawn much attention in the analysis of natural compounds due to its high speed, improved sensitivity, compound specificity and better accuracy. The MS gives the same mass for isomeric and isobaric compounds, and therefore, these cannot be characterized by MS alone unless supported by prior separation. HPLC/UPLC and MS are thus complementary to each other and thus enhance each other's analytical power. We have used HPLC-ESI-QTOF-MS to develop a simple and specific HPLC-ESI-QTOF-MS method for identification and characterization, and UPLC-ESI-QTRAP-MS method for rapid, sensitive and specific quantification of selected major bioactive compounds (gallic acid, protocatechuic acid, caffeic acid, quercetin, ellagic acid, rutin, kaempferol-3-*O*-rutinoside, luteolin, kaempferol, quinic acid and ursolic acid, phyllanthin and hypophyllanthin) in the aerial parts of *P. amarus* collected from three different locations in India. The standards used were obtained from Extrasynthese (Genay, France) and CSIR-CIMAP (Lucknow, India).

2.2 PLANT COLLECTION AND EXTRACTION

The *P. amarus* plant material was collected from three different locations in India – Kolkata, West Bengal (WB); Jabalpur, Madhya Pradesh (MP); and Lucknow, Uttar Pradesh (UP) – in 2009 with herbarium specimens deposited in the medicinal plant herbarium of CSIR-CDRI. All the collections were again repeated in 2010 and 2011.

The shade-dried plant material was ground in a mechanical mill to fine powder. The powdered samples (25 g each) were separately mixed with 100 mL of ethanol, sonicated for 15 min and kept for 48 h at room temperature (25°C). It was filtered using Whatman filter paper. The residue was re-extracted repeatedly six times with fresh ethanol following the same procedure. The combined filtrates were evaporated to dryness under a reduced pressure of 20–50 kPa at 40°C using a Buchi rotary evaporator. 1 mg/mL solution of dried crude plant extract was prepared in methanol and filtered through a 0.22-μm PVDF membrane into an HPLC autosampler vial prior to HPLC-ESI-QTOF-MS/MS analysis.

2.3 LC-MS ANALYSIS OF PHYTOCHEMICALS

The mass spectrometer used for qualitative analysis and structural characterization was an Agilent 6520 hybrid quadrupole time of flight mass spectrometer (Agilent Technologies, Santa Clara, CA, USA) interfaced with an Agilent 1200 HPLC equipped with quaternary pump (G1311A), online vacuum degasser (G1322A), autosampler (G1329A) and a diode-array detector (G1315D). A Thermo Betasil C8 column (250 mm×4.5 mm, 5 μ) operated at 25°C was used for the separation of the components. The mobile phase consisted of 0.1% formic acid in water (A) and methanol (B), and was delivered at a flow rate of 0.4 mL/min under a gradient program (0.1–10 min, 35%–40% B; 10–15 min, 40%–60% B; 15–35 min, 60%–80% B; 35–45 min, 80%–92% B; 45–50 min, 92% B; 50–57 min, 92%–35% B). The sample injection volume was 1.5 μL. The diode-array detector was set to monitor at 254 and 280 nm, and the online UV spectra were recorded in the scanning range of 190–400 nm. The mass spectrometer was operated in negative and positive electrospray ionization modes, and mass spectra were recorded by scanning the mass range of m/z 50–1,500 in both MS and MS/MS modes. Nitrogen was used as drying, nebulizing and collision gas. Flow rate of the drying gas was 12 L/min. The heated capillary temperature was set to 350°C, and nebulizer pressure was set at 45 psi. The source parameters, capillary voltage (VCap), fragmentor, skimmer and octapole voltages were set to 3,500, 175, 65 and 750 V, respectively. For the MS/MS analysis, collision energy was set at 20, 25, 30, 35 and 40 eV. The accurate mass data of the molecular ions were processed through Mass Hunter version B 04.00 (Agilent Technologies).

For quantitative analysis, a hybrid linear ion trap triple-quadrupole mass spectrometer (API 4000 QTRAP™MS/MS system from AB Sciex, Concord, ON, Canada) equipped with electrospray (Turbo V) ion source interfaced with an Acquity ultra-performance liquid chromatography (UPLC) system having an auto sampler and a binary pump (Waters, Milford, MA, USA) and a 10-μL sample loop was used. Separation was achieved on an Acquity CSH C18 column (2.1 mm×100 mm, 1.7 μm) at a temperature of 30°C using a gradient elution of 0.1% (v/v) formic acid–water (A) and acetonitrile (B) at a flow rate of 0.3 mL/min. The gradient program consisted of an initial linear increase from 30% to 70% B over 1.9 min, then an increase to 90% B during 1.9–2.4 min, followed by a hold of 90% B up to 3.6 min and then back to the initial condition by 3.9 min. The volume of sample injection was 2 μL.

The QTRAP mass spectrometer was operated in positive and negative ESI modes with polarity switching. The optimized parameters for positive mode were as follows: the ion spray voltage was set to 5,500 V; the turbo spray temperature, 500°C; nebulizer gas (gas 1), 50 psi; heater gas (gas 2), 50 psi; collision gas, medium; and curtain gas (CUR), 20 psi. The optimized parameters for negative mode were as follows: the ion spray voltage was set to −4200 V; the turbo spray temperature, 550°C; nebulizer gas (gas 1), 20 psi; heater gas (gas 2), 20 psi; collision gas, medium; and curtain gas (CUR),

20 psi. The ESI-MS analysis was carried out in the mass range from m/z 100 to 1,000. Analyst 1.5.1 software package (AB Sciex) was used for instrument control and data acquisition.

2.4 QUALITATIVE ANALYSIS

The extracts of *P. amarus* samples from WB, MP and UP were analyzed on the HPLC-QTOF instrument, and the base peak chromatograms (BPCs) are shown in Figure 2.1. Retention time (t_R), exact mass, molecular formula, MS/MS data and geographical distribution of identified compounds are presented in Table 2.1. The phyto-components are numbered as they appear in the BPC. The compounds are classified into five groups for discussion: (1) hydroxybenzoic and hydroxycinnamic acid derivatives, (2) flavonoids, (3) ellagic acid and their derivatives, (4) lignans and (5) other compounds.

2.4.1 Hydroxybenzoic and Hydroxycinnamic Acid Derivatives

Based on the accurate mass and MS/MS data, 13 compounds were tentatively identified as hydroxybenzoic and hydroxycinnamic acid derivatives. Compounds **1** and **18** were identified as *β*-glucogallin and syringin, respectively. The MS/MS spectra of these two compounds yielded base peaks at m/z 169.0136 $[M\text{-}H\text{-}C_6H_{10}O_5]^-$ for **1** and 209.0812 $[M\text{-}H\text{-}C_6H_{10}O_5]^-$ for **18** due to loss of the sugar moiety and fragments at m/z 211.0246 $[M\text{-}H\text{-}C_4H_8O_4]^-$ 125.0042 $[M\text{-}H\text{-}C_6H_{10}O_5\text{-}CO_2]^-$ for **1** and m/z 194.0046 $[M\text{-}H\text{-}C_6H_{10}O_5\text{-}CH_3]^-$, 148.0546 for **18** (Hossain et al. 2010; Yang et al. 2012). Compounds **3** and **7** were identified as isomeric trigalloylhexose eluting at 6.6 and 8.3 min, respectively. They showed similar product ions at m/z 483.0761 $[M\text{-}H\text{-}C_7H_4O_4]$–, 465.0656 $[M\text{-}H\text{-}C_7H_6O_5]$– and 313.0575 $[M\text{-}H\text{-}C_7H_6O_5\text{-}C_7H_4O_4]$– due to losses of a galloyl group, gallic acid and another galloyl group (Tuominen and Sundman 2013; Yang et al. 2012). Compounds **4** and **6** were identified and characterized as geraniin and castalin and yielded common fragments at m/z 300.999 and 169.014 corresponding to ellagic acid and gallic acid, respectively (Santos et al. 2011; Sudjaroen et al. 2012; Yang et al. 2012). Compounds **5**, **9**, **17** and **20** were identified as gallic acid, protocatechuic acid, caffeic acid and 4-hydroxybenzoic acid, respectively, and the base peaks in their mass spectra were observed at m/z 125.0228, 109.0212, 135.0447 and 93.0351, respectively,

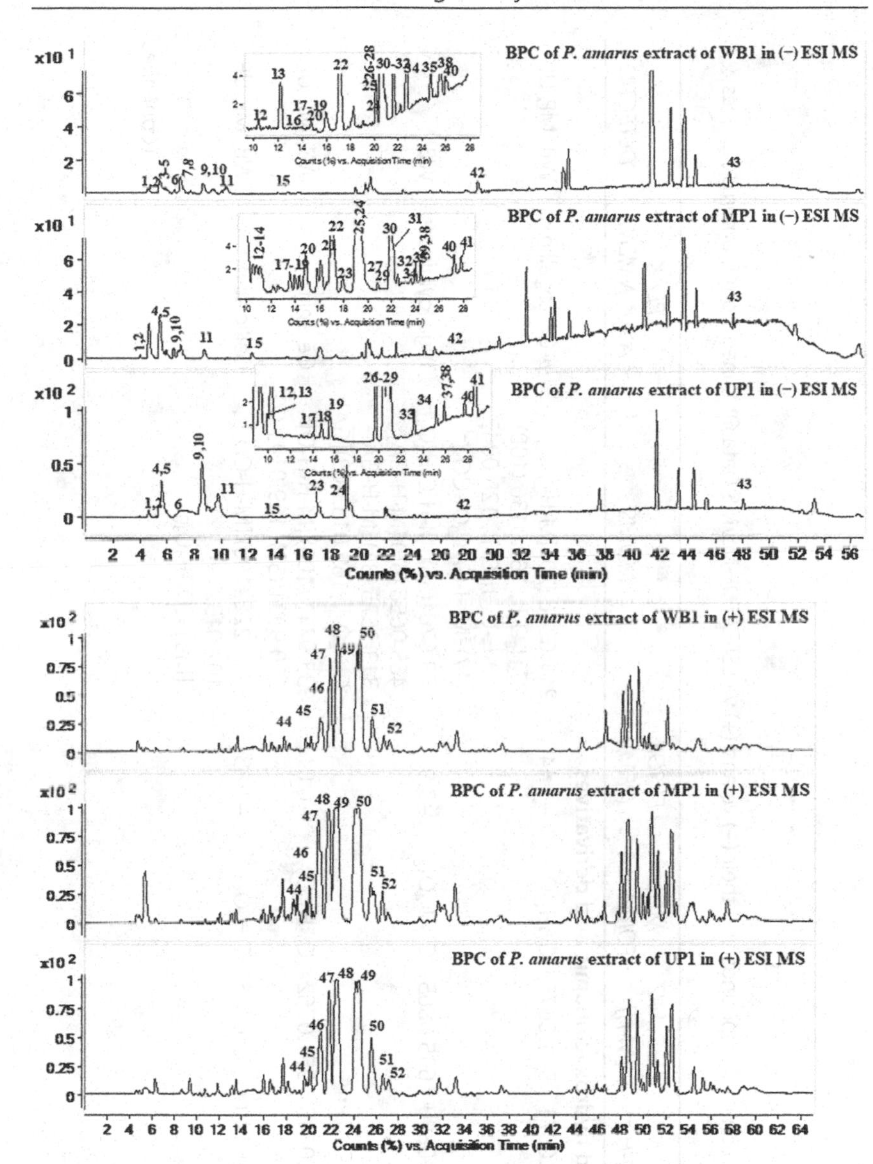

FIGURE 2.1 BPCs of *P. amarus*. WB_1: West Bengal collection 1, MP_1: Madhya Pradesh collection 1 and UP_1: Uttar Pradesh collection 1 in (–) ESI-MS and (+) ESI-MS modes. (Reproduced from Kumar et al. 2015a with permission from Elsevier.)

TABLE 2.1 Identification of Compounds and their (–) ve and (+) ve ESI-Q-TOF-MS/MS Data Obtained from *P. amarus* Samples (WB, MP and UP)

COMP. NO.	*RT (MIN)*	*CAL. M/Z [M-H]⁻*	*OBS. M/Z [M-H]⁻*	*MOL. FORMULA*	*ERROR (PPM)*	*MS/MS*	*PEAK ASSIGNMENT*	*DETECTION*
Hydroxybenzoic and hydroxycinnamic acid derivatives								
1	5.7	331.0671	331.0675	$C_{13}H_{16}O_{10}$	−1.44	211.0246 (20) [M-H-$C_4H_8O_4$]⁻, 169.0136 (100) [M-H-$C_6H_{10}O_5$]⁻, 125.0042 (17) [M-H-$C_6H_{10}O_5$-CO_2]⁻	*β*-Glucogallin	WB, MP, UP
3	6.6	635.0890	635.0865	$C_{27}H_{24}O_{18}$	−3.9	483.0761(36) [M-H-$C_7H_4O_4$]⁻, 465.0656 (65) [M-H-GA]⁻, 313.0575 (31) [M-H-GA⁻-$C_7H_4O_4$]⁻, 169.0133(63)[GA]⁻	Trigalloylglucose	WB
4	6.7	951.0745	951.0762	$C_{41}H_{28}O_{27}$	−0.01	933.0717 (100) [M-H-H_2O]⁻ 300.9991 (52), 169.0141 (2)	Geraniin	WB, MP, UP
5	6.9	169.0140	169.0145	$C_7H_6O_5$	−1.85	125.0228 (100) [M-H-CO_2]⁻, 107.0132 (9) [M-H-CO_2-H_2O]⁻	Gallic acid[a]	WB, MP, UP

(*Continued*)

TABLE 2.1 (*Continued*) Identification of Compounds and their (–) ve and (+) ve ESI-Q-TOF-MS/MS Data Obtained from *P. amarus* Samples (WB, MP and UP)

COMP. NO.	*RT (MIN)*	*CAL. M/Z [M-H]⁻*	*OBS. M/Z [M-H]⁻*	*MOL. FORMULA*	*ERROR (PPM)*	*MS/MS*	*PEAK ASSIGNMENT*	*DETECTION*
6	7.2	631.0577	631.0586	$C_{27}H_{20}O_{18}$	1.24	461.033 (71) [M-H-$C_7H_4O_4$-H_2O]⁻, 445.0461(17) [M-H-$C_7H_4O_5$-H_2O]⁻, 300.9986(78) [ellagic acid]⁻, 273.0030, 245.0092(44), 229.0142(45), 169.0143[GA]⁻ (100), 125.0254 (30)	Castalin	WB, UP
7	8.3	635.0890	635.0891	$C_{27}H_{24}O_{18}$	0.15	483.0761(36) [M-H-$C_7H_4O_4$]⁻,465.0656 (65) [M-H-GA]⁻, 313.0575 (31) [M-H-GA⁻-$C_7H_4O_4$]⁻,169.0133(63)[GA]⁻	Trigalloylglucose (isomer)	WB
8	8.8	633.0733	633.0733	$C_{27}H_{22}O_{18}$	–0.04	431.0618 [M-H-$C_7H_4O_4$]⁻,463 0528 (19) [M-H-$C_7H_4O_4$-H_2O]⁻, 300.9988(100), 275.0201 (18)	Corilagin	WB
9	9.7	153.0193	153.0195	$C_7H_6O_4$	–1.28	109.0276 (100) [M-H-CO_2]⁻, 108.0212 (98) [M-H-CO_2-H]⁻	Protocatechuic acid[a]	WB, MP, UP

(*Continued*)

TABLE 2.1 (*Continued*) Identification of Compounds and their (–) ve and (+) ve ESI-Q-TOF-MS/MS Data Obtained from *P. amarus* Samples (WB, MP and UP)

COMP. NO.	*RT (MIN)*	*CAL. M/Z [M-H]⁻*	*OBS. M/Z [M-H]⁻*	*MOL. FORMULA*	*ERROR (PPM)*	*MS/MS*	*PEAK ASSIGNMENT*	*DETECTION*
11	11.3	183.0299	183.0299	$C_8H_8O_5$	–0.26	125.0158 (6), 124.0162 (100) $[M\text{-}H\text{-}CH_3\text{-}CO_2]^-$	Methyl gallate	WB, MP, UP
13	12.4	163.0395	163.0399	$C_9H_8O_3$	0.99	119.0498 (33) $[M\text{-}H\text{-}CO_2]^-$, 93.0348(100)$[M\text{-}H\text{-}CO_2\text{-}C_2H_2]^-$	*p*-Coumaric acid	WB, MP, UP
17	14.0	179.0344	179.0348	$C_9H_8O_4$	0.87	135.0447(100) $[M\text{-}H\text{-}CO_2]^-$, 117.0344 (6) $[M\text{-}H\text{-}CO_2\text{-}H_2O]^-$	Caffeic acid[a]	WB, MP, UP
18	14.3	371.1348	371.1347	$C_{17}H_{24}O_9$	0.24	209.0812(100) $[M\text{-}H\text{-}C_6H_{10}O_5]^-$,194.0046 (48), $[M\text{-}H\text{-}C_6H_{10}O_5\text{-}CH_3]^-$, 148.0546	Syringin	WB, MP, UP
20	14.7	137.0244	137.0247	$C_7H_6O_3$	–2.26	93.0351 (100) $[M\text{-}H\text{-}CO_2]^-$, 75.0240(5) $[M\text{-}H\text{-}CO_2\text{-}H_2O]^-$	4-Hydroxyben zoic acid	WB, MP, UP
Flavanones								
22	16.6	433.1140	433.1138	$C_{21}H_{22}O_{10}$	0.42	271.0649 (100) $[Y_0]^-$151.0066 (88) $[^{1,3}A_0]^-$, 119.0499 (17) $[^{1,3}B_0]^-$	Prunin	WB, MP, UP
24	19.8	433.1140	433.1146	$C_{21}H_{22}O_{10}$	–1.35	271.0605 (100) $[Y_0]^-$,151.0038 (88) $[^{1,3}A_0]^-$, 119.0499 (17) $[^{1,3}B_0]^-$	Prunin (isomer)	WB, MP, UP
30	21.4	287.0561	287.0561	$C_{15}H_{12}O_6$	–1.89	151.0025 (15) $[^{1,3}A_0]^-$, 125.0235 (100)	Eriodictyol	MP, UP

(*Continued*)

TABLE 2.1 (*Continued*) Identification of Compounds and their (–) ve and (+) ve ESI-Q-TOF-MS/MS Data Obtained from *P. amarus* Samples (WB, MP and UP)

COMP. NO.	*RT (MIN)*	*CAL. M/Z [M-H]⁻*	*OBS. M/Z [M-H]⁻*	*MOL. FORMULA*	*ERROR (PPM)*	*MS/MS*	*PEAK ASSIGNMENT*	*DETECTION*
38	25.3	271.0612	271.0612	$C_{15}H_{12}O_5$	–0.58	151.0074 (43) $[^{1,3}A_0]^-$, 119.0537 (100) $[^{1,3}B_0]^-$	Naringenin	UP
Flavonols								
21	15.8	625.1414	625.1416	$C_{27}H_{30}O_{17}$	0.31	301.0389 (100) $[^{3,7}Y_0]^-$, 271.0264 (7) $[^{3,7}Y_0\text{-CH2O}]^-$, 151.0043 (6) $[^{1,3}A_0]^-$	Quercetin-3,4′-di-*O*-glucoside[a]	WB, MP, UP
26	20.1	463.0877	463.0863	$C_{21}H_{20}O_{12}$	–0.36	301.0257 (50) $[Y_0]^-$,300 0257(100) $[Y_0\text{-H}]^-$, 271.0226 (19) $[Y_0\text{-CH}_2\text{O}]^-$, 255.0299 (11) $[Y_0\text{-CO-H}_2\text{O}]^-$, 151 0000 $[^{1,3}A_0]^-$	Quercetin-*O*-hexoside	WB, MP
27	20.7	609.1461	609.1453	$C_{27}H_{30}O_{16}$	–1.3	301.0354(96) $[Y_0]^-$, 300.02153(100) $[Y_0\text{-H}]^-$ 271.0245(9) $[Y_0\text{-CH}_2\text{O}]^-$, 151.0017 $[^{1,3}A_0]^-$	Rutin[a]	WB, UP
28	21.1	463.0877	463.0897	$C_{21}H_{20}O_{12}$	–0.39	301.0257 (50) $[Y_0]^-$, 300 0257(100) $[Y_0\text{-H}]^{-\bullet}$, 271.0226 (19) $[Y_0\text{-H}_2\text{CO}]^-$, 255.0299 (11) $[Y_0\text{-CO-H}_2\text{O}]^-$, 151 0000 $[^{1,3}A_0]^-$	Quercetin-*O*-hexoside (isomer)	WB, MP, UP

(*Continued*)

TABLE 2.1 (*Continued*) Identification of Compounds and their (–) ve and (+) ve ESI-Q-TOF-MS/MS Data Obtained from *P. amarus* Samples (WB, MP and UP)

COMP. NO.	*RT (MIN)*	*CAL. M/Z [M-H]⁻*	*OBS. M/Z [M-H]⁻*	*MOL. FORMULA*	*ERROR (PPM)*	*MS/MS*	*PEAK ASSIGNMENT*	*DETECTION*
29	21.3	463.0882	463.0887	$C_{21}H_{20}O_{12}$	–1.1	316.0211(100) $[Y_0\text{-H}]^-$ 300.0280 (3) $[Z\text{-H}]^-$, 271.0243 (16) $[Y_0\text{-H}_2CO_2]^-$ 255.0996(8)$[Z\text{-H}_2CO_2]$,151$[^{1,3}A_0]^-$	Myricitrin	WB, MP, UP
31	21.8	447.0933	447.0936	$C_{21}H_{20}O_{11}$	–0.63	300.0280 (100) $[Y_0\text{-H}]^{-\bullet}$, 284.0329 (63) $[Z\text{-H}]^-$, 271.0242 (37) $[Y_0\text{-H}_2CO]^-$, 151.0025(16) $[^{1,3}A_0]^-$	Quercitrin	WB, MP, UP
33	22.6	593.1512	593.1513	$C_{27}H_{30}O_{15}$	–0.3	285.0406 (100), 255.0315 (15)	Kaemferol-3-*O*-rutinoside[a]	WB, MP
34	22.9	447.0933	447.0938	$C_{21}H_{20}O_{11}$	–1.19	285.0413 (100) $[Y_0]^-$,	Fisetin-*O*-hexoside	WB, MP
37	24.5	431.0984	431.0985	$C_{21}H_{20}O_{10}$	–0.83	285.0365 (30) $[Y_0]^-$, 255.0299 (100) $[Y_0\text{-CH}_2O]^-$, 227.0358 (64)	Kaempferol-*O*-hexoside	WB, MP, UP
39	25.4	301.0354	301.0358	$C_{15}H_{10}O_7$	–1.52	151.0034 (100) $[^{1,3}A_0]^-$, 107.0131 (58) $[^{1,3}A_0\text{-CO}_2]^-$ or $[^{1,3}B_0^-\text{-CH}_2O_2]^-$	Quercetin[a]	WB, MP, UP
40	25.03	285.0405	285.0407	$C_{15}H_{10}O_6$	–0.81	151.0075 (23) $[^{1,3}A_0]^-$ 133.0320 (100) $[^{1,3}B_0]^-$	Kaempferol[a]	MP

(*Continued*)

TABLE 2.1 (*Continued*) Identification of Compounds and their (–) ve and (+) ve ESI-Q-TOF-MS/MS Data Obtained from *P. amarus* Samples (WB, MP and UP)

COMP. NO.	*RT (MIN)*	*CAL. M/Z [M-H]⁻*	*OBS. M/Z [M-H]⁻*	*MOL. FORMULA*	*ERROR (PPM)*	*MS/MS*	*PEAK ASSIGNMENT*	*DETECTION*
Flavones								
32	22.4	593.1506	593.1507	$C_{27}H_{30}O_{15}$	0.84	593.1516 (4) $[M-H]^-$, 285.0403 (100) $[Y_0]^-$, 255.0280 (10) $[Y_0-CH_2O]^-$	Luteolin-*O*-dihexoside	WB, MP, UP
35	23.0	447.0927	447.0938	$C_{21}H_{20}O_{11}$	−1.19	285.0413 (100) $[Y_0]^-$ 255.0348 (66) $[Y_0-CH_2O]^-$	Luteolin-*O*-hexoside	WB, MP, UP
41	26.7	285.0405	285.0413	$C_{15}H_{10}O_6$	3.1	175.0418 (21), 151.0074 (23), 133.0320 (100), 121.0334 (8), 107.0168 (25)	Luteolin[a]	WB, MP, UP
Ellagic acid and Derivatives								
12	11.5	463.0518	463.0522	$C_{20}H_{16}O_{13}$	0.9	300.9984 (100) $[M-H-C_6H_{13}O_5]^-$, 283.9962) $[M-H-C_6H_{10}O_5-H_2O]^-$ (1), 257.0035 (6) $[M-H-C_6H_{10}O_5-CO_2]^-$, 244.0031(5), 169.0126(5), 125.0253 (17)	Ellagic acid-*O*-hexoside	WB, MP, UP
16	13.7	329.0303	329.0296	$C_{16}H_{10}O_8$	0.03	314.0069 (15) $[M-H-CH_3]^-$, 298.9836 (100) $[M-H-2CH_3]^-$, 270.9892 (92) $[M-H-2CH_3-CO]^-$	Di-*O*-methyl Ellagic acid	WB

(*Continued*)

TABLE 2.1 (*Continued*) Identification of Compounds and their (–) ve and (+) ve ESI-Q-TOF-MS/MS Data Obtained from *P. amarus* Samples (WB, MP and UP)

COMP. NO.	*RT (MIN)*	*CAL. M/Z [M-H]⁻*	*OBS. M/Z [M-H]⁻*	*MOL. FORMULA*	*ERROR (PPM)*	*MS/MS*	*PEAK ASSIGNMENT*	*DETECTION*
25	20.0	300.9990	300.9998	$C_{14}H_6O_8$	–2.74	283.9958 (18) $[M-H-OH]^-$, 257.0087 (8) $[M-H-CO_2]^-$	Ellagic acid[a]	WB, MP, UP
42	27.5	343.0459	343.0463	$C_{17}H_{12}O_8$	0.12	328.0259 (17) $[M-H-CH_3]^-$, 312.9999 (100) $[M-H-2CH_3]^-$, 297.9737 (64) $[M-H-3CH_3]^-$ 269.9803 (19) $[M-H-3CH_3-CO]^-$	Tri-*O*-methylellagic acid	MP, UP
Other compounds								
2	5.8	191.0561	191.0561	$C_7H_{12}O_6$	–0.03	127.04 $[M-H-CO-2H_2O]^-$, 109.03, 93.04, 85.03 (100)	Quinic acid	WB, MP, UP
10	9.9	291.0146	291.0145	$C_{13}H_8O_8$	0.43	247.0252 (100) $[M-H-CO_2]^-$, 219.0301 (17) $[M-H-CO_2-CO]^-$, 191.0343 (51) $[M-H-CO_2-2CO]^-$, 173.0251 (51) $[M-H-CO_2-2CO-H_2O]^-$, 145.0299, 119.0510	Brevifolin carboxylic acid	WB, MP, UP

(*Continued*)

TABLE 2.1 (*Continued*) Identification of Compounds and their (–) ve and (+) ve ESI-Q-TOF-MS/MS Data Obtained from *P. amarus* Samples (WB, MP and UP)

COMP. NO.	*RT (MIN)*	*CAL. M/Z [M-H]⁻*	*OBS. M/Z [M-H]⁻*	*MOL. FORMULA*	*ERROR (PPM)*	*MS/MS*	*PEAK ASSIGNMENT*	*DETECTION*
14	13.0	435.1291	435.1298	$C_{21}H_{24}O_{10}$	−0.45	273.0766(42) [M-H-$C_6H_{10}O_5$]$^-$, 179.0387(9), 167.0351 (100), 123.0452(16)	Phloridzin	WB, MP, UP
15	13.5	387.1661	387.1663	$C_{18}H_{28}O_9$	−0.66	207.1023 (6) [M-H-$C_6H_{10}O_5$-H_2O]$^-$, 59.0159 (100)	Tuberonic acid hexoside	WB, MP, UP
19	14.5	305.0303	305.0303	$C_{14}H_{10}O_8$	0.03	273.0036(43) [M-H-CH_3OH]$^-$, 245.0082 (80) [M-H-CH_3OH-CO]$^-$, 217.0139 (100) [M-H-CH_3OH-2CO]$^-$, 201.0184 (15) [M-H-CH_3OH-CO-CO_2]$^-$, 189.0188 (23) [M-H-CH_3OH-3CO]$^-$, 173.0243(15) [M-H-CH_3OH-2CO-CO_2]$^-$	Methylbrevifolin carboxylate	WB, MP, UP

(*Continued*)

TABLE 2.1 (*Continued*) Identification of Compounds and their (–) ve and (+) ve ESI-Q-TOF-MS/MS Data Obtained from *P. amarus* Samples (WB, MP and UP)

COMP. NO.	*RT (MIN)*	*CAL. M/Z [M-H]⁻*	*OBS. M/Z [M-H]⁻*	*MOL. FORMULA*	*ERROR (PPM)*	*MS/MS*	*PEAK ASSIGNMENT*	*DETECTION*
23	19.6	319.0459	319.0459	$C_{15}H_{12}O_8$	0.13	273.0047(6) $[M-H-CH_3OCH_3]^-$, 245.0082 (80) $[M-H-CH_3OCH_3-CO]^-$, 217.0139 (100) $[M-H-CH_3OCH_3-CO-CO]^-$, 201.0184 (15) $[M-H-CH_3OCH_3-CO-CO_2]^-$,189.0188 (23) $[M-H-CH_3OCH_3-3CO]^-$, 173.0243(15) $[M-H-CH_3OCH_3-2CO-CO_2]^-$	Dimethylbrevifolin carboxylate	WB, MP, UP
36	24.3	357.1344	357.1359	$C_{20}H_{22}O_6$	−0.11	342.1318 (7) $[M-H-CH_3]^-$, 311.1016	Pinoresinol	WB, MP,
43	48.3	455.3531	455.353	$C_{30}H_{48}O_3$	−0.04	407.314 (5) $[M-HCHO-H_2O]^-$	Ursolic acid[a]	WB, MP, UP

(+) ve ESI-MS/MS data of unknown lignans obtained from aerial part of *P. amarus*. (Reproduced from Kumar et al. 2015a with permission from Elsevier)

(*Continued*)

TABLE 2.1 (*Contniued*) Identification of Compounds and their (–) ve and (+) ve ESI-Q-TOF-MS/MS Data Obtained from *P. amarus* Samples (WB, MP and UP)

COMP NO.	*RT*	*OBS M/Z [M+H]+*	*ERROR (PPM)*	*MOL. FORMULA*	*MS/MS*	*ASSIGNMENT*	*DISTRIBUTION*
46	22.1	436.2687	0.39	$C_{23}H_{30}O_5$	355.1919 (30), 340.1682 (47), 323.1669 (2), 217.1225 (5), 203.1071 (3), 191.1069 (6), 177.0912 (14), 165.0905 (19), 151.0753 (100), 147.0797 (7), 137.0593 (5), 121.0642 (1)	Phyllanthin 1	WB, MP, UP
48	23.1	355.1904	0.03	$C_{22}H_{26}O_4$	240.1666 (32), 217.1204 (4), 203.1067 (2), 191.1060 (5), 177.0907 (16), 151.0742 (100), 147.0791 (5), 137.0585 (6), 136.0506 (6), 121.0646 (1)	Phyllanthin 2	WB, MP, UP

(Continued)

TABLE 2.1 (*Contniued*) Identification of Compounds and their (–) ve and (+) ve ESI-Q-TOF-MS/MS Data Obtained from *P. amarus* Samples (WB, MP and UP)

COMP NO.	*RT*	*OBS M/Z [M+H]⁺*	*ERROR (PPM)*	*MOL. FORMULA*	*MS/MS*	*ASSIGNMENT*	*DISTRIBUTION*
49	23.7	387.2168	0.69	$C_{23}H_{30}O_5$	355.1934 (3), 340.1708 (45), 223.1634 (1), 217.1240 (5), 203.1079 (5), 191.1080 (8), 177.0923 (23) 165.0921 (20), 151.0800 (100), 147.0812 (8), 137.0602 (6), 136.0528 (10), 121.0655 (2)	Phyllanthin 3	WB, MP, UP
51	24.5	401.1952	1.7	$C_{23}H_{28}O_6$	354.1440 (4), 203.1041 (2), 191.0692 (6), 177.0902 (48), 165.0538 (34), 151.0746 (100), 147.0793 (3), 137.0586 (3), 136.0512 (2), 121.0641 (3)	Phyllanthin 4	WB, MP, UP
52	25.1	369.1686	2.3	$C_{22}H_{24}O_5$	354.1455 (4), 308.1387, 203.1019 (3), 191.0697 (9), 177.0898 (72), 165.0538 (49), 155.0744 (100), 147.0744 (10), 137.0584 (8), 136.0508 (4), 121.0637 (7)	Phyllanthin 5	WB, MP, UP

(+) ve ESI-MS/MS data of known lignans obtained from aerial part of *P. amarus*. (Reproduced from Kumar et al. 2015a with permission from Elsevier)

(*Continued*)

TABLE 2.1 (*Contniued*) Identification of Compounds and their (–) ve and (+) ve ESI-Q-TOF-MS/MS Data Obtained from *P. amarus* Samples (WB, MP and UP)

COMP NO.	T_R	*OBS M/Z [M+H]⁺*	*MS/MS*	*ASSIGNMENT*
44	20.3	355.1176	261.0886 (1), 231.0784 (3), 203.0834 (5), 173.0587 (2), 147.0422 (1), 135.0431 (100), 103.0537 (8)	Hinokinin
45	21.6	453.1881 [M+Na]⁺	420.1725 (7), 370.1706 (37), 352.1702 (37), 307.1255 (21), 210.9999 (20), 177.0882 (28), 166.0506 (26), 163.0677 (34), 151.0700 (100)	Hypophyllanthin
47	22.4	419.2425	355.1911 (32), 340.1673 (48), 323.1634 (2), 309.1485 (1), 217.1207 (5), 203.1054 (3), 191.1055 (5), 177.0893 (14), 165.0898 (17), 151.0745 (100), 147.0788 (9), 137.0574 (5), 121.0625 (1)	Phyllanthin
50	23.8	453.1880 [M+Na]⁺	420.1725 (7), 370.1706 (37), 352.1702 (37), 307.1255 (21), 210.9999 (20), 177.0882 (28), 166.0506 (26), 163.0677 (34), 151.0700 (100)	Hypophyllanthin (isomer)

[a] Matched with reference standards, WB: West Bengal, MP: Madhya Pradesh, UP: Uttar Pradesh, Obs: observed.

Source: Reproduced from Kumar et al. 2015a with permission from Elsevier.

due to loss of CO_2 (Hossain et al. 2010; Iswaldi et al. 2013; Maity et al. 2013; Sudjaroen et al. 2012; Yang et al. 2012). The MS and MS/MS data of gallic acid and protocatechuic acid were also matched with authenticated standards. Compound **8** was identified as corilagin which gave fragments at m/z 481.0618 and 463.0528 due to consecutive losses of dehydrogallic acid and H_2O and at m/z 300.9989 due to ellagic acid moiety (Dincheva et al. 2013; Pfundstein et al. 2010; Sudjaroen et al. 2012). Compound **11** was identified as methyl gallate, since its fragment was observed at m/z 124.0162 as base peak due to simultaneous losses of methyl radical and CO_2 (Pfundstein et al. 2010; Santos et al. 2011; Sudjaroen et al. 2012). Compound **13** was identified as *p*-coumaric acid which shows major fragments at m/z 119.0498 due to loss of CO_2 and m/z 93.0348 (base peak) due to further loss of C_2H_2 (Hossain et al. 2010; Iswaldi et al. 2013; Sánchez-Rabaneda et al. 2003a,b).

2.4.2 Flavonoids

The nomenclature proposed by Domon and Costello for glycoconjugates is used for describing the product ions of flavonoid glycosides (Cuyckens and Claeys 2004; Dincheva et al. 2013). All CID spectra obtained for deprotonated flavonoid-*O*-glycosides showed a radical aglycone ion $[Y_0\text{-H}]^{-\bullet}$ besides the regular $[Y_0]^-$ ion, corresponding to the deprotonated aglycone (Cuyckens and Claeys 2004; Dutra et al. 2014). The $[Y_0\text{-H}]^{-\bullet}$ ion is formed by a homolytic cleavage of the glycosidic bond between the aglycone and the glycan, whereas the $[Y_0]^-$ is formed by heterocyclic cleavage (Cuyckens and Claeys 2004).

Four compounds (**22**, **24**, **30** and **38**) were tentatively identified as flavanones. Compound **22** was identified as prunin which showed characteristic peak at m/z 271.0605 $[Y_0]^-$ due to loss of sugar moiety, and its fragments were observed at m/z 151.0038 $[^{1,3}A_0]^-$ and 119.0499 $[^{1,3}B_0]^-$ *via* retro-Diels–Alder reaction (RDA) (Sánchez-Rabaneda et al. 2003a,b). Compound **24** was assigned as an isomer of prunin with $[M\text{-H}]^-$ at m/z 433.1138 with an elution time of 19.8 min and MS/MS products similar to compound **22**. Compound **30** was identified and characterized as eriodictyol, and its MS/MS spectrum showed the characteristic peak at m/z 151.0025 $[^{1,3}A_0]^-$ *via* RDA (Santos et al. 2011; Yang et al. 2012). Compound **38** was identified and characterized as naringenin, and its MS/MS product ions were observed at m/z 151.0074 $[^{1,3}A_0]^-$ and 119.0537 $[^{1,3}B_0]^-$ due to RDA (Sánchez-Rabaneda et al. 2003a,b; Yang et al. 2012) (Figure 2.2).

Five flavonols (compounds **21**, **26**, **27**, **28** and **31**) were tentatively identified and characterized as quercetin 3,4-*O*-diglucoside, quercetin-*O*-hexoside, rutin, quercetin-*O*-hexoside (isomer) and quercitrin, respectively. They showed

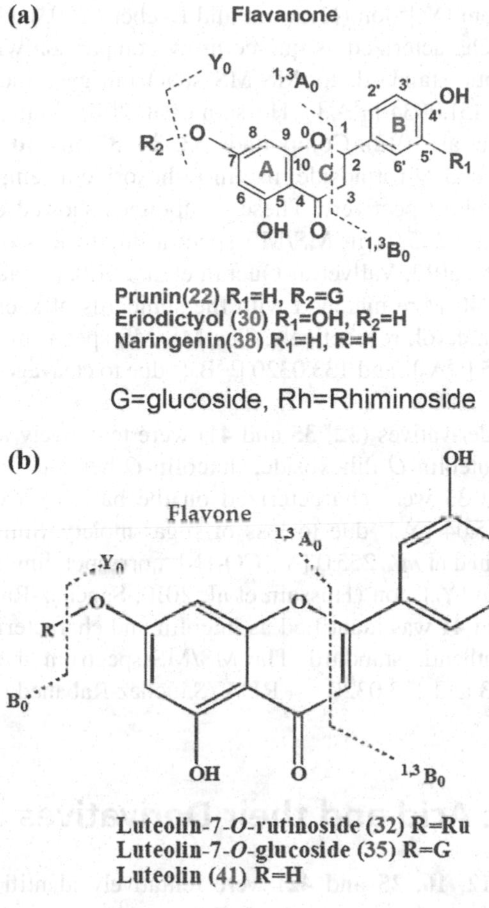

(a) Flavanone

Prunin(22) R_1=H, R_2=G
Eriodictyol (30) R_1=OH, R_2=H
Naringenin(38) R_1=H, R=H

G=glucoside, Rh=Rhiminoside

(b) Flavone

Luteolin-7-*O*-rutinoside (32) R=Ru
Luteolin-7-*O*-glucoside (35) R=G
Luteolin (41) R=H

FIGURE 2.2 Fragmentation of (a) flavanones and (b) flavones.

MS/MS product ions $[Y_0]^-$ and $[Y_0\text{-H}]^{-\bullet}$ due to loss of sugar moiety. The aglycone radical ion $[Y_0\text{-H}]^{-\bullet}$ was observed as a base peak at high collision energy (Cuyckens and Claeys 2004). All these compounds showed common fragment at m/z 271.02 due to loss of CHO / CH_2O from $[Y_0\text{-H}]^{-\bullet}$ / $[Y_0]^-$. They also showed the characteristic fragment at m/z 151.00 $[^{1,3}A_0]^-$ (Fabre et al. 2001; Hvattum and Ekeberg 2003). For authentication, data of compounds **21** and **26** were matched with their standards, quercetin 3,4-*O*-diglucoside and rutin, respectively. Compound **29** was tentatively identified as myricitrin which yielded two fragments at m/z 316.0211 (100) $[Y_0\text{-H}]^-$ and 300.0280 $[Z\text{-H}]^-$ via cleavage of C-O bond between sugar moiety and aglycone at different position. The fragment at m/z 271.0243 $[Y_0\text{-}H_2CO_2]^-$ was observed corresponding to loss of H_2O+CO from $[Y_0]^-$ ion (Hvattum and Ekeberg 2003). Compound **39** was identified and characterized as quercetin by comparison with the retention time of authentic standard. Its MS/MS spectrum gave the characteristic fragment at m/z 151.0034 $[^{1,3}A_0]^-$ (Hossain et al. 2010; Sánchez-Rabaneda et al. 2003a; Yang et al. 2012). Compounds **33**, **34**, **37** and **40** were characterized as kaempferol-3-*O*-rutinoside, fisetin-*O*-hexoside, kaempferol-*O*-hexoside and kaempferol, respectively. These compounds showed characteristic fragment $[Y_0]^-$ at m/z 285.03 in MS/MS spectra due to loss of sugar from $[M\text{-}H]^-$ (Iswaldi et al. 2013; Vallverdú-Queralt et al. 2012). For authentication, compounds **33** and **40** were matched with the standards of kaempferol-3-*O*-rutinoside and kaempferol, respectively. The MS/MS spectrum gave product ions at m/z 151.0075 $[^{1,3}A_0]^-$ and 133.0320 $[^{1,3}B_0]^-$ due to cleavage of ring C *via* RDA (Llorach et al. 2003).

Three flavone derivatives (**32**, **35** and **41**) were tentatively identified and characterized as luteolin-*O*-dihexoside, luteolin-*O*-hexoside and luteolin. Compounds **32** and **35** were characterized on the basis of a characteristic fragment at *m/z* 285.04 $[Y_0]^-$ due to loss of sugar moiety. Similar common fragment was obtained at *m/z* 255.0 $[Y_0\text{-CO-}H_2]^-$corresponding to loss of CO followed by H_2 from $[Y_0]^-$ ion (Hossain et al. 2010; Sánchez-Rabaneda et al. 2003a,b). Compound **41** was identified as luteolin and characterized by comparison with the authentic standard. The MS/MS spectrum showed product ions at *m/z* 151.0073 and 133.0320 *via* RDA (Sánchez-Rabaneda et al. 2003b) (Figure 2.2).

2.4.3 Ellagic Acid and their Derivatives

Four compounds (**12**, **16**, **25** and **42**) were tentatively identified as ellagic acid derivatives. Using reference standard ellagic acid as a template, the compounds were characterized. Compound **25** was identified as ellagic acid by comparison with authentic standard and by the observed MS/MS fragments at m/z 283.9958 $[M\text{-}H\text{-}OH]^-$ and m/z 257.0087 $[M\text{-}H\text{-}CO_2]^-$. Compound **12** was identified as ellagic acid-*O*-hexoside which produced characteristic fragments at m/z 300.9998 by loss of the hexoside moiety and at m/z 283.99 and 257.00 as in ellagic acid (Sudjaroen et al. 2012; Tuominen and Sundman 2013; Yang et al. 2012). Compounds **16** and **42** were identified as di-*O*-methyl ellagic acid and tri-*O*-methyl ellagic acid, respectively. Compound **16** showed fragments at m/z 314.0069 $[M\text{-}H\text{-}CH_3]^-$, 298.9836 $[M\text{-}H\text{-}2CH_3]^-$ and 270.9892 $[M\text{-}H\text{-}2CH_3\text{-}CO]^-$, and compound **42** showed fragments at 328.0259 $[M\text{-}H\text{-}CH_3]^-$, 312.9999 $[M\text{-}H\text{-}2CH_3]^-$, 297.973 $[M\text{-}H\text{-}3CH_3]^-$ and 269.9803 $[M\text{-}H\text{-}3CH_3\text{-}CO]^-$ (Pfundstein et al. 2010).

2.4.4 Lignans

There is much less information available regarding the identification, characterization and quantification of lignans in plants by mass spectrometry (Eklund et al. 2008; Peñalvo et al. 2005). In *P. amarus* extract, four known compounds of phyllanthin series, namely, phyllanthin (**47**), hypophyllanthin (**45**), hypophyllanthin isomer (**50**) and hinokinin (**44**), and one more lignin pinoresinol (**36**) were identified by HPLC-ESI-QTOF-MS in positive ionization mode. Compound **36** (*m/z* 357.1359) was identified as pinoresinol, and the product ions were observed at *m/z* 342.1318 $[M\text{-}H\text{-}CH_3]^-$ and 311.1016 $[M\text{-}H\text{-}CH_3\text{-}CH_3O]^-$ (Eklund et al. 2008). Using reference standard phyllanthin as a template, five unknown lignans (**46, 48, 49, 51** and **52**) were tentatively identified. Hypophyllanthin was detected as $[M+Na]^+$, whereas phyllanthin and hinokinin gave $[M+H]^+$ ions. The exact mass, molecular formula and MS/MS data are given in Table 2.1 for known lignans. The MS/MS spectrum of protonated phyllanthin at m/z 419.2425 showed product ions corresponding to the loss of $2CH_3OH$ followed by CH_3, CH_3OH and CH_3OCH_3 giving rise to the ions at m/z 355, 340, 323 and 309, respectively. Simple bond cleavages gave rise to the other major fragments at m/z 217, 203, 191, 177, 165, 151 and 137. Compound **45** was identified as hypophyllanthin at m/z 453.1880 $[M+Na]^+$. The base peak of hypophyllanthin was observed at m/z 151, corresponding to fragment $[C_9H_{11}O_2]^+$ similar to phyllanthin, and the other fragments are listed in Table 2.1. For authentication, the retention time (t_R 21.6 min) and MS/MS spectrum of **45** were matched with that of standard hypophyllanthin. Compound **44** was tentatively identified and characterized as a hinokinin. The base peak was observed at m/z 135, corresponding to the fragment $[C_8H_7O_2]^+$, and its MS/MS spectrum was matched with that in the previously reported literature (Eklund et al. 2008). Compound **50** was identified as the isomer of hypophyllanthin which had different retention time (t_R 23.8 min), but produced the same fragmentation pattern. The presence of the common fragments in compounds **46, 48, 49, 51** and **52** indicates that the five lignans, *viz*., phyllanthin 1–5, belong to phyllanthin series. These compounds showed common fragments except **51**, which showed two fragments at m/z 354 and 309 instead of 355 and 309 as shown in Table 2.1.

2.4.5 Other Compounds

Compound **2**, corresponding to quinic acid, was identified and characterized with the help of reference standard. The characteristic fragment of quinic acid was observed at m/z 127.04 due to losses of CO and $2H_2O$ (Santos et al. 2011; Karar and Kuhnert 2015). Three compounds (**10, 19** and **23**) were identified as brevifolin derivatives, namely, brevifolin carboxylic acid, methylbrevifolin carboxylate and dimethylbrevifolin carboxylate. The $[M-H]^-$ ion of **10** generated fragment ions at m/z 247.0252 due to loss of CO_2 and further consecutive loss of CO resulted in the ions at m/z 219.0301 and m/z 191.0343 (Tuominen and Sundman 2013). Compounds **19** (m/z 305.0303) and **23** (m/z 319.0459) showed the same fragment ion at m/z 273.0036 due to losses of CH_3OH and CH_3OCH_3, respectively. The fragment ion at m/z 273 successively lost one, two and three molecules of CO resulting in the ions at m/z 245.00, 217.0139 and 189.0188, respectively (Sudjaroen et al. 2012). Compound **14** was characterized as phloridzin. It produced major fragments at *m/z* 273.0766 $[M-H-C_6H_{10}O_5]^-$ due to loss of hexoside moiety and *m/z* 179.0387, 167.0351 and 123.0452 resulting from cleavages of the remaining aglycone unit (Hossain et al. 2010; Hvattum and Ekeberg 2003; Santos et al. 2011). Compound **15** was characterized as a tuberonic acid hexoside, and its major fragments were found at m/z 207.1023 corresponding to loss of hexose and m/z 59.0159 due to carboxylate ion (Amessis-Ouchemoukh et al. 2014). Compound **43** was characterized as ursolic acid with the help of its standard and characteristic fragment ions at m/z 407.3314 corresponding to losses of HCHO and H_2O (Chen et al. 2011; Hu et al. 2018).

2.5 QUANTITATIVE ANALYSIS

There is no reported method for the simultaneous quantification of lignans and flavonoids in *P. amarus*. Hence, we selected phyllanthin and hypophyllanthin, the characteristic lignans of *P. amarus*, along with other bioactive compounds including flavonoids (kaempferol-3-*O*-rutinoside, quercetin) and acids (gallic acid, protocatechuic acid) for simultaneous quantification. Curcumin and piperine were used as internal standards in ESI (+) and ESI (–) modes, respectively, at the appropriate place. This method is not only useful for the quantification of other lignans, flavonoids and acids in this plant but also useful for the assessment of quantitative variations due to geographical changes.

The quantification method was validated for linearity, LOQs and LODs, precision, stability and recovery as per the International Conference on Harmonization Guidelines (Guideline 2005). The satisfactory results obtained are listed in Table 2.2, which shows that the correlation coefficient in the range between 0.9986 and 0.9998 confirms high reproducibility. The results further revealed that the minimum concentration levels at which the analytes could be reliably detected (LODs) and quantified (LOQs) were found to be 0.18–1.82 and 0.44–3.82 ng/mL, respectively, indicating the high sensitivity of this method. The % RSDs of intra- and inter-day precision were less than 1.12% and 1.92%, respectively.

TABLE 2.2 Regression Equation, Correlation Coefficients, Linearity Ranges and Lower LOD and LOQ for Six Reference Analytes

ANALYTES	*REGRESSION EQUATION*	R^2	*LINEAR RANGE (PPB)*	*LOD (NG/ML)*	*LOQ (NG/ML)*	*PRECISION RSD (%)*		*STABILITY RSD % (N = 5)*	*RECOVERY RSD (%)*
						INTRADAY (N = 6)	*INTERDAY (N = 6)*		
Gallic acid	0.0124+0.925x	0.9992	5–250	0.71	1.82	0.84	0.73	2.31	1.57
Protocatechuic acid	0.0136+1.8081x	0.9997	1–50	0.16	0.91	1.03	1.04	3.42	1.33
Kaempferol-3-*O*-rutinoside	0.0046+1.0996x	0.9996	1–100	0.17	0.58	0.52	1.92	1.31	1.66
Quercetin	0.0298+1.1063x	0.9998	1–100	0.13	0.44	0.61	0.31	2.55	2.82
Hypophyllanthin	−0.0401+0.0283x	0.9987	10–250	1.82	3.82	0.93	1.36	1.75	1.91
Phyllanthin	0.0012+0.0148x	0.9986	1–100	0.18	0.52	1.12	1.02	1.73	2.14

Source: Reproduced from Kumar et al. 2015a with permission from Elsevier.

Similarly, % RSDs for stability and recovery for all components were less than 3.42% and 2.82%, respectively. These data demonstrated that the established method was precise, accurate and sensitive enough for simultaneous quantitative determination of all these six compounds in *P. amarus* samples.

2.6 DISTRIBUTION OF BIOACTIVE COMPOUNDS

The developed UPLC-MRM method was subsequently applied to determine the contents of six bioactive compounds, namely, gallic acid, protocatechuic acid, kaempferol-3-*O*-rutinoside, quercetin, phyllanthin and hypophyllanthin in the *P. amarus* samples from WB, MP and UP. Remarkable differences were observed in the contents of these bioactive compounds in all samples as shown in Table 2.3. Phyllanthin and hypophyllanthin were detected in the highest amount in samples WB_1-WB_3 (44.70–54.20 mg/g) and (13.50–23.70 mg/g), respectively, in comparison with MP_1-MP_3 and UP_1-UP_3. Most of the variations in distribution of bioactive compounds were observed in samples MP_1-MP_3 and UP_1-UP_3. Kaempferol-3-*O*-rutinoside was the highest in MP_1 (35 mg/g), followed by quercetin (0.012–5.220 mg/g), protocatechuic acid (0.230–0.541 mg/g) and gallic acid (3.85–32.80 mg/g). Gallic acid, protocatechuic acid, hypophyllanthin and phyllanthin were detected in most of the samples, but the amounts varied between the years of collection. It was clearly observed that MP has the lowest quantities of hypophyllanthin and phyllanthin (MP_2=1.15 mg/g and MP_1=0.85 mg/g, respectively). The difference in contents of bioactive compounds collected either from different geographical locations or in different years might have been influenced by various factors such as climatic conditions, growing environment, age of the plant, time of harvesting, cultivation techniques, process for drying and storage condition (Douglas et al. 2004; Zhang et al. 2011).

2.7 PCA OF HPLC-ESI-QTOF-MS FINGERPRINTS

Principal component analysis (PCA) is a clustering method that reduces the dimensionality of multivariate data to obtain a new set of variables, principal components (PCs). It is an unbiased tool to describe significant data variance

TABLE 2.3 Content of Gallic Acid, Protocatechuic Acid, Kaempferol-3-O-rutinoside, Quercetin, Hypophyllanthin and Phyllanthin from Aerial Parts of *P. amarus*

CONTENT (MG/G)	*GALLIC ACID*	*PROTOCATECHUIC ACID*	*KAEMPFEROL-3-O-RUTINOSIDE*	*QUERCETIN*	*HYPOPHYLLANTHIN*	*PHYLLANTHIN*
WB_1	1.74	0.42	0.03	nd	22.90	54.20
WB_2	0.45	0.24	nd	0.07	13.50	48.00
WB_3	1.99	0.10	0.07	0.14	23.70	44.70
MP_1	0.57	0.54	0.35	0.01	2.07	0.85
MP_2	32.8	0.42	0.01	5.22	1.15	30.10
MP_3	3.85	0.23	0.07	nd	7.61	46.60
UP_1	0.72	0.18	bdl	0.12	5.72	38.00
UP_2	5.01	0.33	nd	nd	16.90	50.50
UP_3	2.06	0.35	bdl	nd	5.35	5.76

bdl: below detection level; nd: not detected.
Source: Reproduced from Kumar et al. 2015a with permission from Elsevier.

in a few PCs. The PCA study of this data set was carried out using the mean values of all the three collections from UP_{1-3}, MP_{1-3} and WB_{1-3} with 43 known peaks which were reduced to 26 peaks in the final PCA model. The information in the peaks of the three regions (MP, UP and WB) was contained in two PCs only. The majority of the variation in the peaks was explained by the first PC, which was able to explain 67.6% variation in data, and the second PC is able to explain 29.1% variation. Thus, both first and second PCs cover 96.7% of peaks information in the studied data set. This is a fairly high amount of information which is able to classify the three regions categorically. On the basis of PC scores, samples MP and WB were closer and lying in the same quadrant, whereas UP was much far apart (Figure 2.3).

In conclusion, rapid and accurate HPLC/ESI-QTOF-MS/MS and UPLC/ESI-MRM methods were established for identification, characterization and quantification of phytoconstituents in the aerial parts of *P. amarus*. Fifty-two compounds including hydroxybenzoic and hydroxycinnamic acid derivatives, flavonoids, ellagic acid derivatives and lignans were unambiguously identified and characterized. Twenty-seven compounds, namely, *β*-glucogallin, trigalloylglucose, trigalloylglucose (isomer), protocatechuic acid, methyl

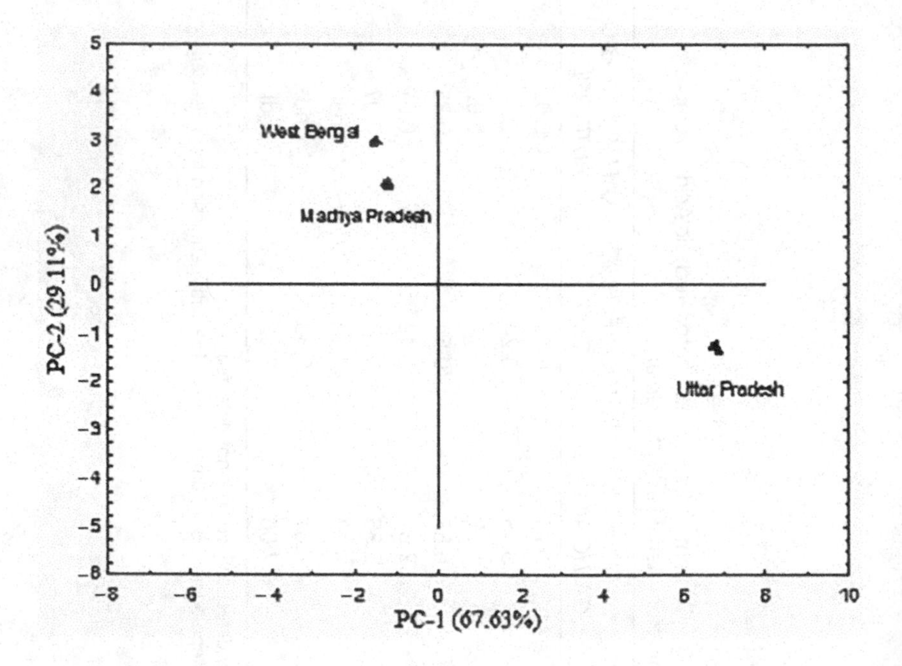

FIGURE 2.3 The PC1 vs. PC2 plot showing the average PC scores for 43 bioactive compounds in the three geographical regions (WB, MP and UP). (Reproduced from Kumar et al. 2015a with permission from Elsevier.)

gallate, *p*-coumaric acid, caffeic acid, syringin, quercetin-diglucoside, kaempferol-*O*-hexoside, kaempferol-3-*O*-rutinoside, kaempferol, quercetin-*O*-hexoside, myricitrin, quercitrin, fisetin-*O*-hexoside, luteolin-*O*-dihexoside, luteolin-*O*-hexoside, luteolin, ellagic acid-*O*-hexoside, di-*O*-methyl ellagic acid, tri-*O*-methylellagic acid, quinic acid, brevifolin carboxylic acid, tuberonic acid hexoside, methylbrevifolin carboxylate and dimethylbrevifolin carboxylate, were detected and characterized for the first time in the aerial parts of *P. amarus* using accurate mass and MS/MS fragmentation pattern. Five unknown lignans, belonging to phyllanthin series (phyllanthin1–5), were tentatively identified based on the fragmentation pattern of phyllanthin standard. The comparative analyses of contents of major bioactive compounds phyllanthin and hypophyllanthin in *P. amarus* obtained from different geographical locations during the period of three years were carried out for the first time by UPLC–ESI-MS/MS. The results clearly indicate variations in the distribution of compounds in different samples of *P. amarus.* PCA revealed that the distribution of bioactive constituents in *P. amarus* from different geographical locations was closer for MP and WB, whereas UP was much far apart.

3 Quantitative Analysis of Bioactive Compounds in *Phyllanthus* Species and its Herbal Formulations

3.1 RATIONALE FOR THE INVESTIGATION

As *Phyllanthus amarus*, *P. niruri*, *P. fraternus* and *P. emblica* contain a large number of bioactive phytoconstituents, they are widely used in herbal formulations (HFs). Although HFs are generally harmless and consumed without any prescription, they may sometimes be not effective or may interact with other drugs. Often, this is due to adulteration or replacement with a similar substitute. It is, therefore, essential that proper assay methods are developed so that individual herbal species can be chemically characterized and quality guaranteed for both herbs and HFs (Chandra et al. 2015). UHPLC-ESI-MS/MS method in MRM acquisition mode is the most effective in rapidly quantifying trace compounds due to its fast separation power, low detection limit, better accuracy, high specificity and resolution (Jin et al. 2017). It was, therefore, decided to develop and validate the UHPLC–ESI-MS/MS method to identify and quantify 23 bioactive compounds, namely, quinic acid (**1**), gallic acid (**2**), protocatechuic acid (**3**), catechin (**4**), chlorogenic acid (**5**), epicatechin (**6**), chebulagic acid (**7**), quercetin-3,4′-di-*O*-glucoside (**8**), chebulinic acid (**9**), ellagic acid (**10**), rutin (**11**), naringin (**12**), kaempferol-3-*O*-rutinoside (**13**), eriodictyol (**14**), scutellarein (**15**), quercetin (**16**), luteolin (**17**), kaempferol (**18**), amentoferine (**19**), hypophyllanthin (**20**) and phyllanthin (**21**), betulinic acid (**22**) and oleanolic acid (**23**) in ethanolic extracts of *P. amarus* whole plant (WP), *P. niruri* (WP), different parts of *P. emblica* and *P. fraternus*, fractions of *P. amarus*, and methanolic extracts of HFs using UHPLC coupled with an ESI hybrid triple quadrupole-linear ion trap mass spectrometer (QqQLIT-MS/MS) in polarity switching MRM mode with the help of reference standards. Principal component analysis (PCA) was used to differentiate the selected *Phyllanthus* species and HFs.

3.2 SAMPLE COLLECTION AND EXTRACTION

Leaves, bark and fruits of *P. emblica* were obtained from CSIR-Indian Institute of Integrated Medicine (CSIR-IIIM), Jammu, India. Leaves, bark and twigs of *P. fraternus* were collected from Aizawl, Mizoram. Certified dried samples of WP of *P. niruri* (Batch No. 10PN-1442) and *P. amarus* (Reference No. PCA/PA/778) were purchased from Tulsi Amrit, Indore, India and Natural Remedies Private Limited, Bangalore, India, respectively. Fourteen different marketed HFs containing *P. amarus*, *P. niruri* and/or *P. emblica* with combination of other medicinal plants purchased from the local market of Lucknow, India were coded as HF1 to HF14. Quantitative variations of selected phytochemicals were detected during the study. Each sample (5 g) was soaked in ethanol (45 mL) followed by sonication at 30°C for 30 min, kept for 48 h and filtered through Whatman filter paper, and the filtrates were concentrated under a reduced pressure of 20–50 kPa at 40°C using a rotary evaporator (Buchi Rotavapor-R2, Flawil, Switzerland). Hexane, butanol and ethyl acetate fractions of *P. amarus* were prepared according to the method suggested by Mojarrab et al. (2013). All extracts and fractions were stored in the refrigerator at –20°C until analysis.

Using a 1 mg/mL stock solution, different concentrations (0.5, 1.0, 1.25, 2.0, 2.5, 5.0, 10.0, 12.5, 20, 25, 50, 100, 125, 200, 250, 500 and 1,000 µg/mL) of standard mixture were prepared for linearity studies. The coating on each HF sample was removed first. The powdered samples (250 mg) were suspended in methanol (25 mL), sonicated for 1 h at 30°C and filtered. The filtrates were diluted with methanol to prepare the final working concentration. The sample was filtered through 0.22-µm syringe filter (Millex-GV, PVDF, Merck Millipore, Darmstadt, Germany) before analysis.

3.3 LC-MS ANALYSIS

The UPLC-ESI-MS/MS analysis was performed on a Waters Acquity UPLC™ system (Waters, Milford, MA, USA) interfaced with a hybrid linear ion trap triple-quadrupole mass spectrometer API 4000 QTRAP™ MS/MS system (AB Sciex, Concord, ON, Canada) equipped with an ESI (Turbo V) source. The Waters Acquity UPLC™ system was equipped with a binary solvent manager, sample manager, column oven and photodiode array detector. The QTRAP mass spectrometer was operated in positive and negative ESI modes with polarity switching. AB Sciex Analyst software version 1.5.1 was used for data acquisition and processing. ACQUITY UPLC BEH™ C18 column (1.7 µm, 2.1 mm × 50 mm) was used for liquid chromatographic separation of targeted compounds at 25°C. Formic acid (0.1%) (A) and acetonitrile (B) were used as the mobile phase with the flow rate of 0.25 mL/min under a gradient program: 0%–5% (B) initial to 1.0 min, 5%–30% (B) from 1 to 5 min, 30%–36% (B) from 5.0 to 7.0 min, 36%–50% (B) from 7.0 to 14.0 min, 50%–90% (B) from 14.0 to 19.0 min, 90%–90% (B) from 19.0 to 20.0 min, 90%–5% (B) from 20 to 22 min, and back to initial conditions in 1 min. The volume of sample injection was 2.0 µL.

MS and MS/MS analyses were carried out in both negative (analytes **1–19** and **22–23**) and positive (analytes **20–21**) ionization modes from m/z 100 to 1,000. Source-dependent parameters, ion spray voltage, temperature (TEM), nebulizer gas (GS 1), heater gas (GS 2) and curtain gas were set at 5500 V, 550°C, 50 psi, 50 psi and 20 psi, respectively. Nitrogen was used as the collision-activated dissociation gas and set as medium, and the interface heater was on. To optimize mass spectrometric conditions, 50 ng/mL solutions in methanol were used at the rate of 10 µL/min by direct infusion using

TABLE 3.1 Compound-Dependent MRM Parameters for *Phyllanthus* Species

PEAK NO.	*RT (MIN)*	*ANALYTE*	*PRECURSOR ION (M/Z)*	*DP (V)*	*EP (V)*	*CE (EV)*	*CXP*	*POLARITY*
1	0.6	Quinic acid	191.0 [M-H]⁻	−67	−12	−33	−13	−
2	1.8	Gallic acid	168.7 [M-H]⁻	−59	−10	−22	−11	−
3	2.9	Protocatechuic acid	152.8 [M-H]⁻	−64	−5	−22	−9	−
4	4.1	Catechin	288.9 [M-H]⁻	−110	−10	−29	−8	−
5	4.4	Chlorogenic acid	353.3 [M-H]⁻	−60	−10	−30	−10	−
6	5.1	Epicatechin	288.6 [M-H]⁻	−120	−9	−29	−9	−
7	5.9	Chebulagic acid	953.1 [M-H]⁻	−176	−10	−45	−16	−
8	6.4	Quercetin-3,4′-di-*O*-glucoside	625.0 [M-H]⁻	−97	−6	−43	−7	−
9	9.1	Chebulinic acid	955.4 [M-H]⁻	−182	−12	−48.9	−20	−
10	9.3	Ellagic acid	300.0 [M-H]⁻	−62	−4	−56	−12	−
11	9.4	Rutin	609.4 [M-H]⁻	−197	−10	−50	−25	−
12	9.9	Naringin	579.1 [M-H]⁻	−131	−9	−45	−9	−
13	11.2	Kaempferol-3-*O*-rutinoside	593.2 [M-H]⁻	−105	−5	−45	−6	−
14	11.3	Eriodictyol	287.0 [M-H]⁻	−88	−8.7	−21	−9	−
15	12.1	Scutellarein	284.8 [M-H]⁻	−99	−5.3	−42	−11	−
16	12.8	Quercetin	301.0 [M-H]⁻	−107	−9	−34	−12	−
17	13.6	Luteolin	285.1 [M-H]⁻	−139	−10	−41	−21	−
18	14.7	Kaempferol	285.0 [M-H]⁻	−95	−5	−49	−10	−
19	16.7	Amentoflavone	537.1 [M-H]⁻	−145	−7	−42	−20	−
20	17.4	Hypophyllanthin	453.4 [M+Na]⁺	97	10	+33	15	+
21	17.6	Phyllanthin	419.1 [M+H]⁺	40	10	+15	10	+
22	19.2	Betulinic acid	455.2 [M-H]⁻	−140	−9	−7	−15	−
23	19.4	Oleanolic acid	455.3 [M-H]⁻	−120	−8	−12	−7	−

Source: Reproduced from Kumar et al. (2017c) with permission from John Wiley and Sons.

a Harvard "22" syringe pump (Harvard Apparatus, South Natick, MA, USA). The most abundant fragment ions in the MS/MS spectrum were selected for MRM transitions. Compound-dependent MRM parameters are shown in Table 3.1. Contents of bioactive compounds obtained from three measurements ($n = 3$) of all samples were used for PCA on STATISTICA, Windows version 7.0 (Stat Soft, USA).

3.4 QUANTITATIVE ANALYSIS OF BIOACTIVE COMPOUNDS

Compounds **1–19**, **22** and **23** (acids, phenols, flavonoids, glycosides and triterpenoids) were detected as $[M-H]^-$at *m/z* 191.0, 168.7, 152.8, 288.9, 353.3, 288.6, 953.1, 625.0, 955.4, 300.0, 609.4, 579.1, 593.2, 287.0, 284.4, 301.0, 285.1, 285.0, 537.1, 455.2 and 455.3, respectively, in negative ionization mode, whereas the lignans **20** and **21** were detected as $[M + Na]^+$ at *m/z* 453.4 and $[M+ H]^+$ at *m/z* 419.1, respectively, in positive ionization mode.

Compounds **1–23** showed the most abundant ions at m/z 84.9, 124.8, 109.3, 203.1, 191.4, 203.0, 300.6, 301.0, 336.9, 145.0, 300.0, 270.9, 285.2, 150.8, 116.9, 150.9, 132.9, 158.9, 375.0, 453.4, 355.2, 455.2 and 455.3 in the MS/MS spectra and were selected as quantifiers.

For optimization of the chromatographic conditions, two types of reverse phase columns, Acquity UPLC CSH™ C18 column (2.1 mm × 100 mm, 1.7 µm) and Acquity UPLC BEH™ C18 column (2.1 mm × 50 mm, 1.7 µm), were tested, but satisfactory separation was achieved on an Acquity UPLC BEH C18 column (2.1 mm × 50 mm, 1.7 µm). Similarly, different concentrations (0.05%, 0.1% and 0.2%) of formic acid were tested to get better ionization, and 0.1% formic acid with acetonitrile gave satisfactory separation. Curtain gas, GS 1 and GS 2 and ion source temperature were also optimized to obtain the highest abundance of precursor and product ions. The different column temperatures (20°C, 25°C, 30°C and 35°C) and flow rates (0.2, 0.25, 0.3, 0.35 and 0.40 mL/min) were also examined. Satisfactory resolution in the shortest analysis time was obtained at a flow rate of 0.25 mL/min at 25°C column temperature. Hence, 0.1% formic acid with acetonitrile at the flow rate of 0.25 mL/min at 25°C on Acquity UPLC BEH™ C18 column (2.1 mm × 50 mm, 1.7 µm) was finally selected for developing the method. MRM total ion chromatograms and extracted ion chromatogram of the analytes are shown in Figure 3.1.

The developed UHPLC-ESI-MS/MS method was validated according to the guidelines of the International Conference on Harmonization (ICH, Q2R1) using determination of calibration curves, lower LOD, lower LOQ, precision, solution stability and recovery (Guideline 2005). Calibration curves were constructed using the peak area of analytes against the corresponding concentrations. Correlation coefficient ($R2 \geq 0.996$) was within the test range, which indicated appropriate correlations between peak areas and concentrations of analytes. The LODs and LOQs for all the analytes were 0.05–1.62 and 0.15–4.95 ng/mL, respectively (Table 3.2).

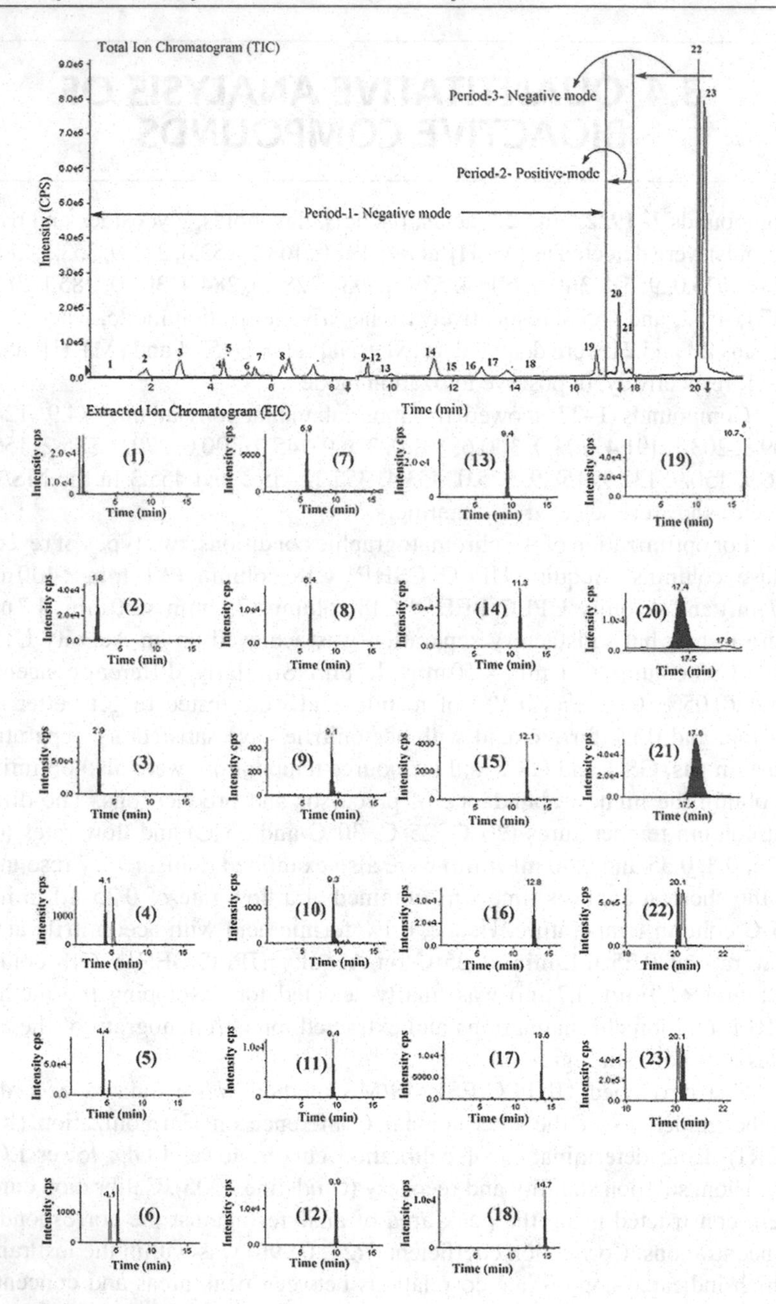

FIGURE 3.1 MRM total ion and extracted ion chromatogram. (Reproduced from Kumar et al. (2017c) with permission from John Wiley and Sons.)

TABLE 3.2 Regression Equation, Correlation Coefficients, Linearity Ranges, LOD, LOQ, Intra- and Interday Precisions, Stability and Recovery for 23 Analytes

PEAK NO.	REGRESSION EQUATION	R^2	LINEAR RANGE (PPB)	LOD (NG/ML)	LOQ (NG/ML)	PRECISION (%RSD) INTRADAY (N = 6)	PRECISION (%RSD) INTERDAY (N = 6)	STABILITY %RSD	RECOVERY MEAN	RECOVERY %RSD
1	$y = 317x + 17$	0.998	0.5–200	0.18	0.54	1.22	1.05	2.45	102.12	1.02
2	$y = 3300x - 174$	1.000	1–100	0.17	0.53	0.21	0.80	1.44	100.22	1.24
3	$y = 5610x + 856$	1.000	2.5–500	0.50	1.54	2.10	1.31	0.45	98.23	1.45
4	$y = 140x - 67.3$	0.999	5–500	1.44	4.41	1.07	0.90	1.21	102.12	1.64
5	$y = 4110x + 907$	0.999	2.5–25	0.73	2.23	1.45	1.23	1.25	100.12	0.95
6	$y = 64.2x + 28.5$	0.996	5–1000	1.46	4.48	1.23	1.45	2.56	99.21	0.99
7	$y = 24800x - 371$	0.998	1–200	0.05	0.15	0.12	0.42	1.92	103.12	2.56
8	$y = 885x + 478$	0.999	5–25	1.62	4.95	0.02	0.09	2.45	102.12	1.84
9	$y = 42.6x - 6.36$	1.000	5–500	0.49	1.51	1.14	1.23	2.41	100.13	1.74
10	$y = 119x + 61.8$	1.000	5–250	1.56	4.76	0.12	0.74	1.89	104.45	1.56
11	$y = 156x - 81.6$	0.999	5–200	1.57	4.80	0.45	2.91	1.25	103.54	2.11
12	$y = 781x - 60.3$	0.999	1–200	0.25	0.78	2.11	1.89	1.87	102.78	1.99
13	$y = 944x + 14.9$	0.999	1–200	0.05	0.16	1.24	1.46	1.45	100.24	2.11
14	$y = 514x + 116$	1.000	2.5–100	0.74	2.28	1.02	1.41	2.56	99.12	2.31
15	$y = 166x - 20$	0.999	2.5–125	0.40	1.22	0.89	1.77	1.58	99.45	2.45
16	$y = 2880x + 521$	0.999	2.5–125	0.60	1.82	0.99	0.19	1.43	100.89	1.23

(Continued)

TABLE 3.2 (*Continued*) Regression Equation, Correlation Coefficients, Linearity Ranges, LOD, LOQ, Intra- and Interday Precisions, Stability and Recovery for 23 Analytes

				LOD	*LOQ*	*PRECISION (%RSD)*			*RECOVERY*	
PEAK NO.	*REGRESSION EQUATION*	R^2	*LINEAR RANGE (PPB)*	*(NG/ML)*		*INTRADAY (N = 6)*	*INTERDAY (N = 6)*	*STABILITY %RSD*	*MEAN*	*%RSD*
17	$y = 1000x + 31.3$	0.999	1–100	0.10	0.32	1.23	1.75	1.45	101.74	1.23
18	$y = 371x + 51$	1.000	2.5–200	0.45	1.39	1.27	1.78	2.45	100.73	1.26
19	$y = 3050x + 148$	0.999	0.5–100	0.16	0.49	0.12	0.81	2.00	100.42	0.78
20	$y = 1020x + 6310$	1.000	2.5–100	0.53	1.62	1.45	1.45	2.13	101.31	0.42
21	$y = 1400x + 172$	1.000	1.5–250	0.41	1.24	1.66	1.21	1.23	103.42	2.64
22	$y = 16300x - 562$	1.000	1–200	0.11	0.35	1.44	0.87	1.25	100.21	2.71
23	$y = 1640x + 469$	0.999	5–250	0.94	2.88	0.78	0.66	2.25	101.45	1.29

Source: Reproduced from Kumar et al. (2017c) with permission from John Wiley and Sons.

The intra- and inter-day precisions were performed with the 23 analytes in six replicates during a single day and triplicate for three consecutive days. The % RSDs for intra- and inter-day precisions obtained were from 0.02% to 2.11% and 0.09% to 2.91%, respectively. Replicate injections of the samples at 0, 2, 4, 8, 12 and 24 h were used at room temperature to investigate the stability of sample solutions. The % RSD values for these repeats were found between 0.45% and 2.56%. The average recoveries were determined by spiking accurately known amounts of all the analytes (high, middle and low) into the sample, which were then extracted and analyzed using formula: recovery = $(a - b)/c \times 100\%$, where a is the detected amount, b is the original amount and c is the spiked amount. The recovery of the method was in the range of 98.22%–104.48% with %RSD ≤ 2.93%, indicating the accuracy of established method (Table 3.2).

3.5 QUANTIFICATION OF THE 23 COMPONENTS

All compounds (**1–23**) were unambiguously identified, characterized and quantified in the extracts and HFs using the developed and validated UPLC-ESI-MS/MS method and the contents presented in Tables 3.3 and 3.4.

The highest contents of hypophyllanthin (20.95 mg/g) and phyllanthin (46.16 mg/g) were detected in ethanolic extracts of *P. amarus* whole plant (WP) followed by *P. niruri* WP (hypophyllanthin: 15.22 mg/g and phyllanthin: 24.12 mg/g). Except for phyllanthin (1.04 mg/g) in twigs of *P. fraternus*, lignans were not detected in the *P. fraternus* samples. Gallic acid (106.00 mg/g), ellagic acid (56.70 mg/g) and quinic acid (21.00 mg/g) were predominant in the fruits of *P. emblica*. However, the leaves of *P. emblica* showed high content of chebulinic acid (32.20 mg/g). In the fractions of *P. amarus* (WP) extract, the highest contents of hypophyllanthin (29.40 mg/g) and phyllanthin (56.60 mg/g) were detected in the ethyl acetate fraction followed by hexane fraction (hypophyllanthin: 14.50 mg/g and phyllanthin: 41.20 mg/g) and butanol fraction (hypophyllanthin: 4.16 mg/g and phyllanthin: 5.89 mg/g). In extract, the maximum contents of gallic acid (10.10 mg/g) and chebulinic acid (1.88 mg/g) were observed in the ethyl acetate fraction. The glycosides, rutin (24.70 mg/g) and kaempferol-3-*O*-rutinoside (2.13 mg/g) were detected high in the butanol fraction of *P. amarus* (WP). The contents of ellagic acid (17.00 mg/g), chebulagic acid (3.00 mg/g) and catechin (0.20 mg/g) were detected in the butanol and ethyl acetate fractions of *P. amarus* WP. Tables 3.3 and 3.4 show the contents (mg/g ± %RSD) of the 23 bioactive compounds in the ethanolic extracts and fractions of *P. amarus* and the four *Phyllanthus* species and HFs.

TABLE 3.3 Contents (mg/g ± %RSD) of 23 Bioactive Compounds in the Ethanolic Extracts of the Four Selected *Phyllanthus* Species and Fractions of *P. amarus*

S. NO.	*PA*	*PN*	*PEB*	*PEL*	*PEF*	*PFT*	*PFS*	*PFL*	*BUT*	*ETY*	*HEX*
1.	0.47.12 ± 0.73	0.23 ± 0.12	0.72 ± 0.17	0.34 ± 0.05	21 ± 0.58	nd	1.10 ± 0.08	18.6 ± 0.71	5.12 ± 0.01	0.89 ± 0.02	nd
2.	1.86 ± 0.22	0.35 ± 0.14	0.45 ± 0.45	6.03 ± 0.69	106 ± 1.00	0.08 ± 0.47	0.06 ± 0.60	0.15 ± 0.61	3.25 ± 0.01	10.1 ± 0.44	nd
3.	0.18 ± 0.52	1.07 ± 0.09	0.01 ± 0.06	0.045 ± 0.12	0.01 ± 0.06	0.19 ± 0.61	0.07 ± 0.51	0.28 ± 0.01	0.19 ± 0.42	0.46 ± 0.1	nd
4.	0.13 ± 0.22	0.18 ± 0.02	nd	nd	0.11 ± 0.19	0.31 ± 0.12	0.25 ± 0.26	nd	0.2 ± 0.06	0.24 ± 0.01	nd
5.	nd	nd	0.003 ± 05	0.02 ± 0.01	nd	nd	nd	0.005 ± 1.02	0.04 ± 0.01	0.02 ± 0.01	nd
6.	nd	0.1 ± 0.06	0.53 ± 0.01	nd	0.05 ± 0.01	nd	nd	0.15 ± 0.01	nd	0.147 ± 0.54	nd
7.	2.12 ± 0.41	0.796 ± 1.60	0.97 ± 0.07	32.2 ± 0.68	24.9 ± 0.96	nd	0.96 ± 0.01	0.73 ± 0.09	3.09 ± 0.38	3.6 ± 0.3	nd
8.	0.08 ± 0.01	0.04 ± 0.26	0.1 ± 0.06	0.05 ± 0.54	0.03 ± 0.56	0.08 ± 0.01	0.06 ± 0.04	0.05 ± 0.01	0.09 ± 0.05	0.1 ± 0.61	nd
9.	nd	1.64 ± 0.24	nd	1.1 ± 0.46	nd	1.64 ± 0.20	1.64 ± 0.32	nd	nd	1.88 ± 0.37	nd
10.	3.35 ± 0.07	0.75 ± 0.05	0.957 ± 0.03	23.3 ± 0.42	56.7 ± 0.32	0.21 ± 0.14	0.21 ± 0.07	0.18 ± 0.02	17.3 ± 0.09	18.5 ± 0.23	nd
11.	1.31 ± 0.15	4.37 ± 0.04	0.368 ± 0.02	0.29 ± 0.35	0.19 ± 0.08	0.64 ± 0.05	0.17 ± 0.42	0.37 ± 0.03	24.7 ± 0.15	1.75 ± 0.01	nd
12.	0.027 ± 0.09	0.012 ± 0.06	0.027 ± 0.02	nd	0.016 ± 0.01	0.016 ± 0.04	0.027 ± 0.03	0.027 ± 0.03	0.027 ± 0.06	0.035 ± 0.06	nd
13.	0.031 ± 0.43	0.318 ± 0.27	nd	nd	nd	0.028 ± 0.02	0.028 ± 0.02	0.054 ± 0.02	2.13 ± 0.41	0.327 ± 0.28	nd
14.	nd	0.004 ± 0.23	nd	nd	nd	nd	nd	nd	nd	0.013 ± 0.02	nd
15.	0.09 ± 0.41	0.15 ± 0.16	0.11 ± 0.42	nd	0.13 ± 0.16	nd	0.11 ± 0.19	nd	0.24 ± 0.06	0.146 ± 0.03	nd

(*Continued*)

TABLE 3.3 (*Continued*) Contents (mg/g ± %RSD) of 23 Bioactive Compounds in the Ethanolic Extracts of the Four Selected *Phyllanthus* Species and Fractions of *P. amarus*

S. NO.	*PA*	*PN*	*PEB*	*PEL*	*PEF*	*PFT*	*PFS*	*PFL*	*BUT*	*ETY*	*HEX*
16.	0.05 ± 0.21	0.12 ± 0.33	0.01 ± 0.25	0.12 ± 0.19	0.14 ± 0.16	0.03 ± 0.07	0.02 ± 0.01	0.03 ± 0.23	0.19 ± 0.4	0.59 ± 0.11	Nd
17.	0.015 ± 0.12	0.012 ± 0.06	0.003 ± 0.04	nd	0.003 ± 0.02	0.012 ± 0.25	0.006 ± 0.02	nd	0.052 ± 0.80	0.031 ± 07	nd
18.	nd	0.05 ± 0.04	nd	0.07 ± 0.22	0.07 ± 0.22	0.001 ± 0.03	0.02 ± 0.07	0.31 ± 0.08	0.43 ± 0.19	0.56 ± 0.06	nd
19.	0.02 ± 0.25	nd	nd	nd	nd	nd	nd	nd	0.04 ± 0.4	0.74 ± 0.17	nd
20.	20.95 ± 0.06	15.22 ± 0.30	0.321 ± 0.03	nd	nd	nd	nd	nd	4.16 ± 0.1	29.4 ± 0.08	14.5 ± 0.12
21.	46.16 ± 0.09	24.12 ± 0.53	nd	nd	0.002 ± 0.22	1.04 ± 0.89	nd	nd	5.89 ± 1.36	56.6 ± 0.55	41.2 ± 0.63
22.	0.08 ± 0.74	0.12 ± 0.07	0.16 ± 0.04	0.04 ± 0.04	0.06 ± 0.12	0.06 ± 0.19	0.14 ± 1.84	0.08 ± 0.06	0.09 ± 0.24	0.07 ± 0.39	0.07 ± 0.24
23.	0.06 ± 0.03	1.52 ± 0.15	0.06 ± 0.10	0.1 ± 1.01	nd	0.12 ± 0.19	0.18 ± 0.13	0.08 ± 0.03	0.02 ± 0.24	0.04 ± 0.21	0.19 ± 1.23

S: Serial; No: Number; PA: *Phyllanthus amarus* (whole plant); PN: *Phyllanthus niruri* (whole plant); PEB: *Phyllanthus emblica* (bark); PEL: *Phyllanthus emblica* (leaves); PEF: *Phyllanthus emblica* (fruits); PFT: *Phyllanthus fraternus* (twigs); PFS: *Phyllanthus fraternus* (stem); PFL: *Phyllanthus fraternus* (leaves); BUT: *P. amarus* butanol fraction; ETY: *P. amarus* Ethyl acetate fraction; HEX: *P. amarus* hexane fraction.
Source: Reproduced from Kumar et al. (2017c) with permission from John Wiley and Sons.

TABLE 3.4 Contents (mg/g ± %RSD) of 23 Bioactive Compounds in Methanolic Extracts of HFs

ANALYTES	*HF1*	*HF2*	*HF3*	*HF4*	*HF5*	*HF6*	*HF7*	*HF8*	*HF9*	*HF10*	*HF11*	*HF12*	*HF13*	*HF14*
1.	1.56 ± 1.08	27.5 ± 1.46	31.20 ± 1.0	16.4 ± 2.31	46.1 ± 2.65	27.20 ± 2.0	85.80 ± 2.31	11.30 ± 3.46	100 ± 2.65	10.40 ± 0.58	10.4 ± 1.15	5.49 ± 0.11	62.80 ± 0.58	11.70 ± 1.00
2.	49.3 ± 0.58	31.6 ± 1.15	27.7 ± 1.00	18.1 ± 1.00	55.5 ± 1.00	125 ± 2.89	12.60 ± 0.58	35.60 ± 2.00	8.90 ± 1.00	22.40 ± 1.00	25.4 ± 4.62	47 ± 0.71	3.87 ± 0.15	30.10 ± 3.21
3.	1.22 ± 1.00	0.701 ± 0.45	0.24 ± 0.17	3.42 ± 2.71	0.97 ± 0.27	1.51 ± 0.37	0.46 ± 0.22	2.62 ± 0.54	1.58 ± 0.50	0.24 ± 0.46	0.25 ± 0.06	1.59 ± 1.85	0.29 ± 0.03	0.67 ± 0.25
4.	0.68 ± 0.23	0.08 ± 0.01	nd	0.05 ± 0.06	0.05 ± 0.19	0.02 ± 0.17	0.03 ± 0.02	0.08 ± 0.05	0.06 ± 0.11	0.04 ± 0.02	0.03 ± 0.11	0.08 ± 0.19	0.14 ± 0.13	0.01 ± 0.57
5.	0.09 ± 0.01	0.20 ± 0.01	0.78 ± 0.57	0.03 ± 0.01	0.37 ± 0.02	1.19 ± 0.10	3.20 ± 0.58	2.05 ± 0.06	0.09 ± 0.01	0.80 ± 0.01	0.05 ± 0.01	0.06 ± 0.01	055 ± 0.03	2.23 ± 0.06
6.	0.13 ± 0.01	nd	0.15 ± 0.01	0.14 ± 0.12	0.04 ± 0.01	0.03 ± 0.01	0.07 ± 0.01	0.02 ± 0.02	0.08 ± 0.01	0.06 ± 0.01	0.03 ± 0.02	0.26 ± 0.02	0.10 ± 0.01	0.06 ± 0.01
7.	0.14 ± 0.29	5.46 ± 0.25	0.10 ± 0.04	0.12 ± 0.12	9.99 ± 0.38	130 ± 68.16	0.31 ± 0.18	0.27 ± 0.05	0.14 ± 0.23	0.13 ± 0.14	0.09 ± 0.03	222 ± 2.52	0.19 ± 0.02	68.40 ± 1.53
8.	0.09 ± 0.01	0.14 ± 1.08	0.075 ± 0.53	0.02 ± 0.06	0.03 ± 0.01	0.03 ± 0.01	0.01 ± 0.51	0.02 ± 0.01	0.07 ± 1.12	0.09 ± 0.01	0.07 ± 1.12	0.08 ± 0.01	0.16 ± 0.48	0.2 ± 0.12
9.	5.51 ± 0.45	17.1 ± 1.08	1.64 ± 0.24	nd	18.4 ± 2.09	255 ± 1.64	2.6 ± 0.35	nd	nd	nd	nd	272.00 ± 1.28	2.60 ± 0.89	51.70 ± 1.89
10.	12.5 ± 1.73	8.93 ± 0.30	3.52 ± 0.05	0.41 ± 0.19	12.5 ± 1.53	41.20 ± 1.00	2.58 ± 0.05	6.22 ± 0.35	14.30 ± 1.53	3.36 ± 0.37	0.222 ± 0.39	55.20 ± 0.58	0.26 ± 0.58	27.60 ± 1.73
11.	0.136 ± 0.08	0.84 ± 0.16	2.69 ± 0.39	0.16 ± 0.18	3.85 ± 0.21	0.19 ± 0.05	3.35 ± 0.21	7.55 ± 0.21	0.03 ± 0.06	1.32 ± 0.07	0.039 ± 0.04	0.376 ± 0.03	0.49 ± 0.02	0.57 ± 0.29
12.	0.071 ± 0.28	0.016 ± 0.00	0.031 ± 0.01	0.012 ± 0.14	0.039 ± 0.01	0.031 ± 0.12	0.016 ± 0.50	nd	0.01 ± 0.04	0.02 ± 0.50	nd	0.04 ± 0.03	nd	0.02 ± 0.12

(Continued)

TABLE 3.4 (*Continued*) Contents (mg/g ± %RSD) of 23 Bioactive Compounds in Methanolic Extracts of HFs

ANALYTES	*HF1*	*HF2*	*HF3*	*HF4*	*HF5*	*HF6*	*HF7*	*HF8*	*HF9*	*HF10*	*HF11*	*HF12*	*HF13*	*HF14*
13.	0.04 ± 0.01	0.08 ± 0.02	0.334 ± 0.31	0.02 ± 0.01	0.40 ± 0.05	0.015 ± 0.02	0.42 ± 0.58	0.92 ± 0.58	0.02 ± 0.02	0.12 ± 0.42	0.002 ± 0.13	0.012 ± 0.01	0.11 ± 0.37	0.18 ± 0.36
14.	0.09 ± 0.06	0.04 ± 0.23	1.00 ± 0.41	nd	nd	0.077 ± 0.02	0.02 ± 0.04	0.02 ± 0.01	0.127 ± 0.06	0.059 ± 0.01	nd	0.09 ± 0.06	0.04 ± 0.05	0.03 ± 0.06
15.	0.01 ± 0.16	0.01 ± 0.17	0.01 ± 0.17	0.01 ± 0.18	0.01 ± 0.40	0.06 ± 0.01	0.04 ± 0.01	0.02 ± 0.13	0.01 ± 0.17	0.05 ± 0.17	nd	0.01 ± 0.17	0.03 ± 0.02	nd
16.	0.50 ± 0.28	0.39 ± 0.25	0.014 ± 0.16	0.002 ± 0.24	0057 ± 0.07	0.026 ± 0.11	0.036 ± 0.03	0.048 ± 0.04	0.01 ± 0.20	0.013 ± 0.03	nd	0.069 ± 0.05	nd	0.005 ± 1.03
17.	0.02 ± 0.18	0.04 ± 0.13	0.10 ± 0.33	0.02 ± 0.18	0.009 ± 0.38	0.40 ± 0.33	0.22 ± 0.05	0.01 ± 0.46	0.12 ± 0.11	0.25 ± 0.18	0.002 ± 0.34	0.202 ± 0.48	0.02 ± 0.14	0.92 ± 0.31
18.	0.084 ± 0.23	0.16 ± 0.14	0.001 ± 0.03	0.03 ± 0.41	0.077 ± 0.18	0.029 ± 0.35	0.031 ± 0.25	0.013 ± 0.14	2.53 ± 0.06	0.032 ± 0.07	nd	0.015 ± 0.17	0.001 ± 0.04	0.006 ± 0.34
19.	nd	nd	nd	0.22 ± 0.30	nd	nd	nd	nd	0.77 ± 0.18	nd	nd	3.83 ± 0.18	nd	nd
20.	0.365 ± 0.07	nd	0.02 ± 0.05	nd	nd	0.095 ± 0.03	nd	0.135 ± 0.06	nd	nd	nd	0.038 ± 0.02	nd	nd
21.	nd	0.786 ± 0.01	nd	nd	nd	nd	nd	5.71 ± 1.23	0.008 ± 0.05	nd	nd	nd	nd	nd
22.	0.048 ± 0.04	0.027 ± 0.02	0.109 ± 0.05	0.665 ± 1.00	0.082 ± 0.06	0.0116 ± 0.04	0.06 ± 0.13	0.491 ± 0.09	0.073 ± 0.06	0.115 ± 0.20	0.013 ± 0.06	0.36 ± 0.24	0.159 ± 0.10	0.005 ± 0.03
23.	0.13 ± 0.09	0.10 ± 0.48	0.14 ± 0.13	1.69 ± 0.45	0.075 ± 0.30	0.045 ± 0.17	0.11 ± 0.06	0.826 ± 0.08	0.096 ± 0.49	nd	0.007 ± 0.03	0.256 ± 0.16	0.082 ± 0.35	0.016 ± 0.04

Source: Reproduced from Kumar et al. (2017c) with permission from John Wiley and Sons.
Nd: Not detected

The total content of all compounds was high in the fruits of *P. emblica* (209.41 mg/g), followed by *P. amarus* (WP) (76.97 mg/g) and *P. emblica* (leaves) (63.70 mg/g). Almost 90% of the total content of *P. emblica* fruit was due to gallic acid, chebulagic acid and ellagic acid. In *P. emblica*, the bark had the minimum total content of 4.8 mg/g, whereas the leaves contained 63.75 g/mg. *P. niruri* (WP) had a total content of 51.17 mg/g, much less than that of *P. amarus*, whereas in *P. fraternus*, the maximum content of 21.09 mg/g was in the leaves which also contained 18.6 mg/g of quinic acid. Among the four plants, *P. niruri* had the maximum content (4.37 mg/g) of rutin. Among the *P. amarus* (WP) fractions, ethyl acetate fraction showed the maximum total content (125.52 mg/g), followed by butanol (67.24 mg/g) and hexane (55.96 mg/g). Among HFs, HF1 and HF2 showed 0.50 mg/g and 0.39 mg/g of quercetin, respectively, whereas high content of eriodictyol (1.00 mg/g) was detected in HF3. The maximum content of oleanolic acid (1.690 mg/g) was detected in HF4, whereas HF6 showed predominantly gallic acid (125 mg/g). Kaempferol-3-*O*-rutinoside (0.922 mg/g) and betulinic acid (0.82 mg/g) were maximum in HF8, but HF9 showed high content of quinic acid (100 mg/g), followed by kaempferol (2.53 mg/g). Chebulinic acid (272.00 mg/g), chebulagic acid (22.20 mg/g) and ellagic acid (55.20 mg/g) were observed predominantly in HF12, whereas HF14 showed high contents of chlorogenic acid (2.23 mg/g) and luteolin (0.918 mg/g). HF12 showed the maximum total content (605.43 mg/g), followed by HF6 (582.17 mg/g) and HF14 (193.59 mg/g).

3.6 DIFFERENTIATION OF PLANTS AND PLANT PARTS

PCA was carried out to differentiate *P. amarus* (WP), *P. niruri* (WP), *P. emblica* (bark), *P. emblica* (leaves), *P. emblica* (fruits), *P. fraternus* (twigs), *P. fraternus* (stem), *P. fraternus* (leaves) and 14 HFs based on the contents of 23 compounds. The first two principal components PC1 and PC2 hold 39.28% and 24.37% in *Phyllanthus* species and plant parts, respectively. PCA showed that the whole plants of *P. amarus* and *P. niruri* were close to each other. However, *P. emblica* (bark), *P. emblica* (leaves), *P. emblica* (fruits), *P. fraternus* (twigs) and *P. fraternus* (stem) were falling together in the same quadrate. *P. fraternus* (leaves) was far from all other species as shown in Figure 3.2a. Similarly, among 14 HFs, the first two principal components PC1 and PC2 hold 42.78% and 19.33%, respectively, of the total variance. HF6 and HF12 were falling in the same quadrate and closer to each other, whereas HF14 was far from all

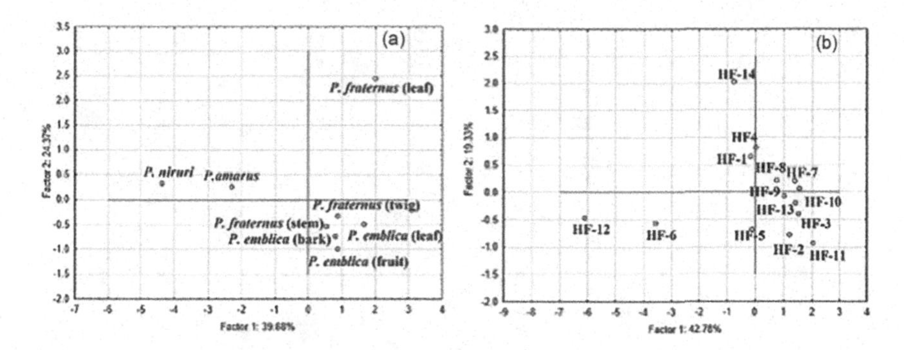

FIGURE 3.2 PCA score plot (a) *P. amarus, P. niruri, P. emblica* (bark), *P. emblica* (leaves), *P. emblica* (fruit), *P. fraternus* (twig), *P. fraternus* (stem), *P. fraternus* (leaves). (b) Herbal formulations. (Reproduced from Kumar et al. (2017c) with permission from John Wiley and Sons.)

others (Figure 3.2b). It is evident from Figure 3.2a that clear differentiation is achieved among the different plants and parts.

In conclusion, the UHPLC-ESI-MS/MS method developed and validated for the simultaneous quantitation of the 23 targeted bioactive compounds in polarity switching MRM mode is very useful in the analysis of *Phyllanthus* plants, parts, extracts and HFs. The total content of 23 compounds was the maximum in the fruits of *P. emblica*, followed by *P. amarus* (WP), *P. niruri* (WP) and *P. fraternus* leaf. Similarly, the maximum contents of hypophyllanthin and phyllanthin were detected in the ethanolic extract of *P. amarus* (WP) and its ethyl acetate and hexane fractions, and hence, these extracts and fractions are very convenient for isolation of phyllanthine and hypophyllanthin. PCA was successfully used to differentiate the *Phyllanthus* species and their HFs.

DART-TOF-MS Analysis of *P. amarus* and Study of Geographical Variations

4

4.1 DIRECT ANALYSIS IN REAL-TIME MASS SPECTROMETRY (DART-MS)

DART is one of the ionization techniques instrumental in moving the sample from the vacuum to the ambient environment (Hajslova et al. 2011). The sample is placed or held in the path of the plasma (containing ions, metastable atoms, electrons and molecules) generated from the DART source using helium or nitrogen. Positive ionization involves atmospheric water cluster ions which transfer H^+ to the thermally desorbed sample, and negative ionization involves electrons reacting with oxygen to produce O_2^- ions which deprotonate the sample to generate $(M\text{-}H)^-$ ions (Cody et al. 2005). DART-MS is now an established ambient ionization technique for the rapid and direct analysis of samples with minimal or no sample preparation without chromatographic separation (Jiang et al. 2010). Since its introduction, it has been widely applied in natural products research (Kumar et al. 2015b; Chernetsova et al. 2012). It has also found effective application in herbal analysis (Wang et al. 2014; Shen et al. 2016). Recently, DART-MS followed by multivariate analysis was used for discrimination of plant parts/species (Kumar et al. 2015b; Singh et al. 2015). DART-TOF-MS combined with principal component analysis (PCA) is an efficient and high-throughput method that allows studying variation in samples. PCA is one of the most widely used multivariate techniques useful in dealing with large data sets for discrimination. PCA can also be used to construct a loadings plot that indicates variables responsible for discrimination among the samples.

As DART-MS fingerprinting is a very fast method, we have used this method for the fingerprinting of *Phyllanthus amarus* samples collected from three different locations in India, viz., Kolkata from the state of West Bengal (WB), Jabalpur from the state of Madhya Pradesh (MP) and Lucknow from the state of Uttar Pradesh (UP) during the period of 2009–2011. The collected plants were washed, dried according to Maity et al. (2013), and powdered.

4.2 DART-MS ANALYSIS

The DART-MS data were obtained on a JMS-100 TLC AccuTOF (Jeol, Tokyo, Japan) mass spectrometer having a DART ion source. The mass spectrometer was operated in positive ion mode with a resolving power of 6000 (full width at half maximum). The orifice 1 potential was set to 28 V, resulting in minimal fragmentation. The ring lens and orifice 2 potentials were set to 13 and 5 V, respectively. Orifice 1 was set at 100°C. The RF ion guide potential was 300 V. The DART ion source was operated with helium gas flowing at approximately 4.0 L/min. The gas heater was set to 300°C. The potential on the discharge needle electrode of the DART source was set to 3,000 V; electrode 1 was 100 V; and the grid was at 250 V. Powdered plant samples were positioned in the gap between the DART source and mass spectrometer with the help of a glass capillary. Data acquisition was from m/z 50 to 1,000, and 15 measurements were made for each sample. Exact mass calibration was accomplished by including a mass spectrum of neat polyethylene glycol (PEG) (1:1 mixture of PEG 200 and PEG 600) in the data file. The mass calibration was accurate to within ±0.002 u. Using the Mass Center software, the elemental composition could be determined on selected peaks.

4.3 CHEMICAL FINGERPRINTS

The representative chemical fingerprints of *P. amarus* samples WB, MP and UP collected in 2009 are shown in Figure 4.1. Though similar ions were present in the DART-MS fingerprints of these samples, their relative abundances were different in each sample. Sixteen compounds were tentatively identified from chemical fingerprints on the basis of their measured exact masses, calculated molecular formula and literature reports. Peaks **1, 2, 3** and **4,** identified as phenazine (m/z 181), norsecurinine (m/z 204), securinine (m/z 218) and

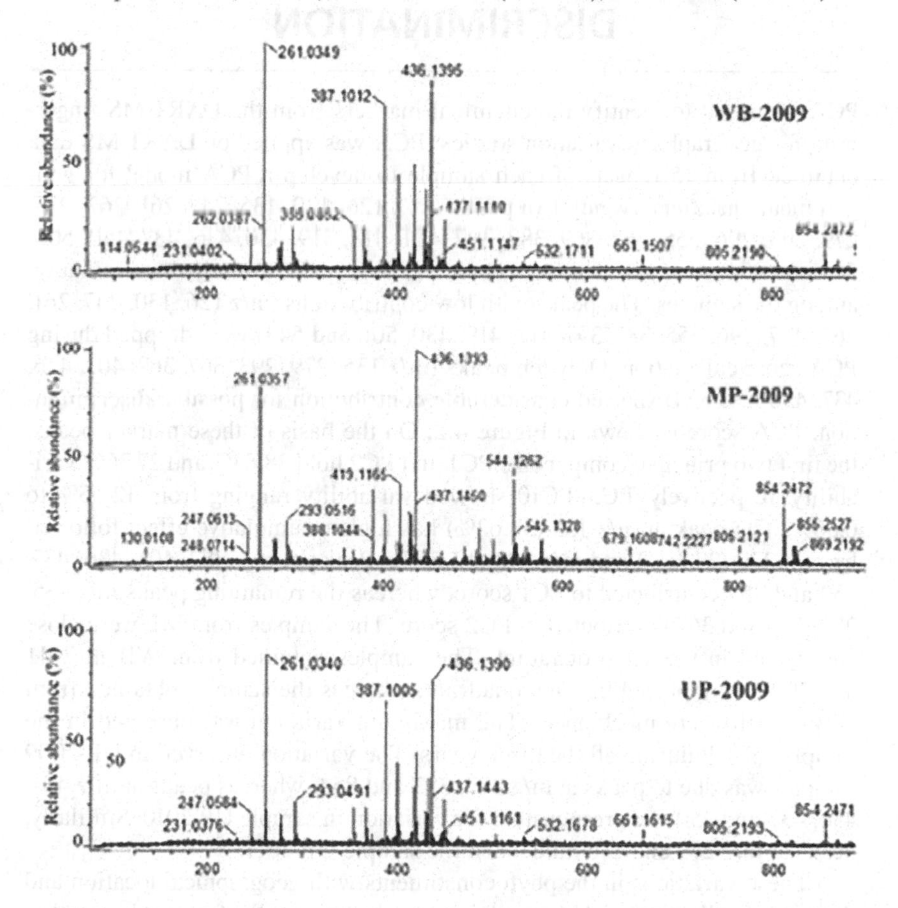

FIGURE 4.1 Representative DART-MS spectra of *P. amarus* samples WB, MP and UP collected in 2009. (Reproduced from Kumar et al. (2017d) with permission from BrJAC.)

dihydrosecurinine (m/z 220), respectively, were alkaloids. Peaks **5–16** were identified as lignans. Peak **14** and **15** were identified as phyllanthin (m/z 419) and hypophyllanthin (m/z 431), respectively. Phyllanthin and hypophyllanthin were detected in all the samples of WB. The m/z 261, 387 and 436 peaks were abundant but could not be identified due to insufficient data and literature reports.

4.4 PRINCIPAL COMPONENT ANALYSIS: DISCRIMINATION

PCA was used to identify the chemical markers from the DART-MS fingerprint for geographical variation studies. PCA was applied on DART-MS data obtained from 15 repeats of each sample to develop a PCA model for geographical variation. Twenty-five peaks (m/z 126, 130, 135, 247, 261, 262, 277, 279, 293, 296, 355, 367, 369, 387, 397, 401, 418, 419, 430, 436, 437, 450, 506, 544 and 854) were selected from chemical fingerprint to study discrimination among the samples. The peaks with low contributions (m/z 126, 130, 247, 261, 262, 277, 296, 355, 387, 397, 418, 419, 430, 506 and 544) were dropped during PCA score calculation. Only ten peaks (m/z 135, 279, 293, 367, 369, 401, 436, 437, 450 and 854) showed considerable contribution for possible discrimination. PCA score is shown in Figure 4.2. On the basis of these marker peaks, the first two principal components PC1 and PC2 hold 38.63% and 27.36% variability, respectively. PC3–PC10 showed variability ranging from 12.55% to 0.66%. The peak at m/z 135 (38.62%) has higher cumulative effect followed by m/z 279 (27.35%) and m/z 293 (12.55%). Peaks at m/z 387, 401, 436, 437, 450 and 854 contributed to PC1 score, whereas the remaining peaks m/z 135, 279, 293 and 367 contributed to PC2 score. The samples from MP were close and placed in the same quadrant. The samples obtained from WB in 2011 and 2009 were placed in same quadrant, whereas the samples obtained from WB in 2010 were much apart. The maximum variation was observed in the samples of UP during all the three years. The variation observed in UP-2009 samples was due to peaks at m/z 279, 387 and 854, whereas peaks at m/z 401, 436, 437 and 450 were responsible for variation in sample UP-2010. Similarly, peaks at m/z 279 and 436 influenced the sample UP-2011.

These variations in the phytoconstituents with geographical location and the year of collection could be explained on the basis of the fact that biosynthesis of secondary metabolites depends on different causes such as attack of herbivores and environmental conditions (Treutter 2006; Figueiredo et al. 2008).

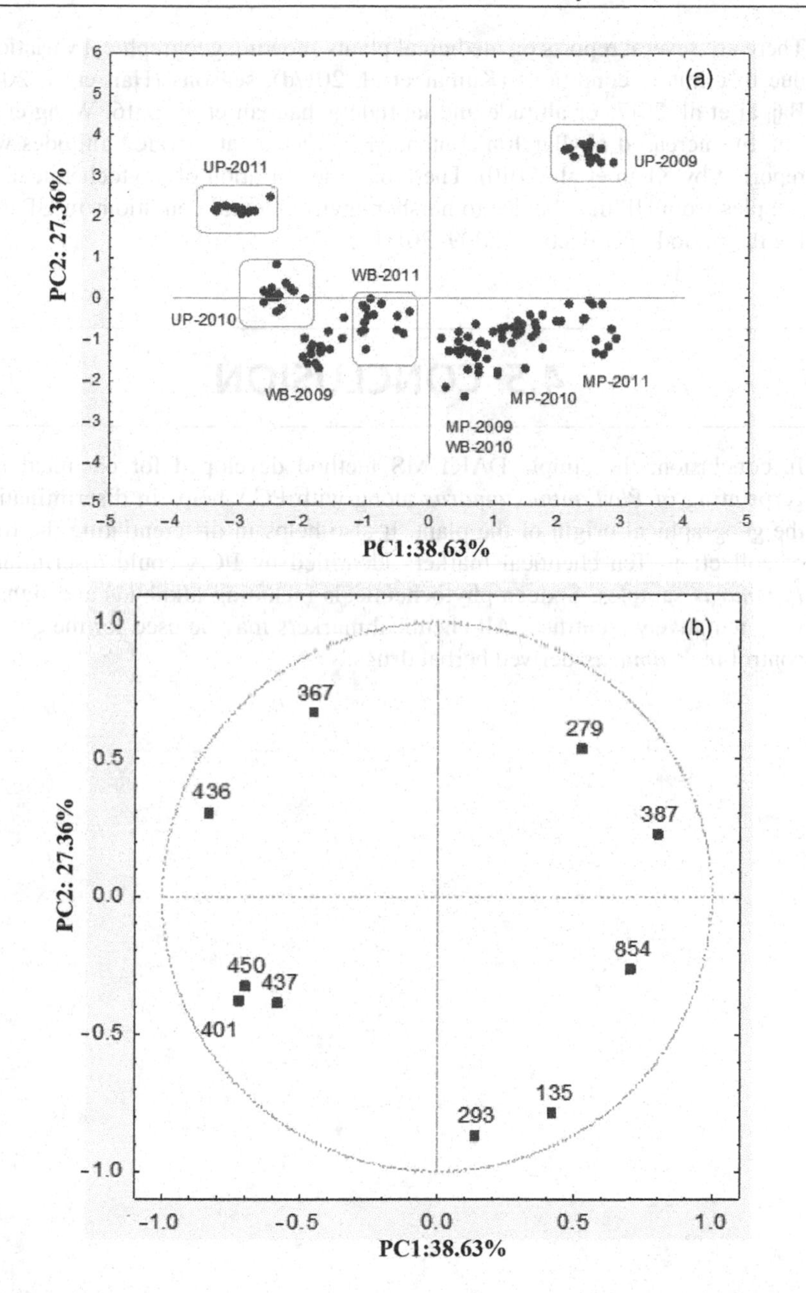

FIGURE 4.2 (A) Scores plot of ten peaks and (B) Loadings plot of ten marker peaks from PCA of DART-MS fingerprint of *P. amarus*. (Reproduced from Kumar et al. (2017d) with permission from BrJAC.)

There are several reports on medicinal plants showing geographical variations due to climatic conditions (Kumar et al. 2017d), seasons (Han et al. 2018; Bajpai et al. 2017) or altitude and latitude (Chauhan et al. 2016; Wang et al. 2017b). Increased phyllanthin content in *P. amarus* at elevated altitudes was reported by Khan et al. (2010). Therefore, the variation of phytochemicals in samples from UP may be due to harsher environmental conditions in UP during the period of collection (2009–2011).

4.5 CONCLUSION

In conclusion, the simple DART-MS method developed for chemical fingerprinting of *Phyllanthus amarus* along with PCA helps in discriminating the geographical origin of the plant. It also helps in differentiating the time of collection. Ten chemical markers identified by PCA could discriminate *P. amarus* samples. Sixteen phytochemicals including alkaloids and lignans were tentatively identified. All chemical markers may be used for the quality control of *P. amarus*-derived herbal drugs.

References

Agharkar, S. P. "Medicinal Plants of Bombay Presidency." Jodhpur, India: Scientific Publishers, 1991.

Al Zarzour, Raghdaa, Mariam Ahmad, Mohd Asmawi, Gurjeet Kaur, Mohammed Saeed, Majed Al-Mansoub, Sultan Saghir, Nasiba Usman, Dhamraa Al-Dulaimi, and Mun Yam. "Phyllanthus Niruri standardized extract alleviates the progression of non-alcoholic fatty liver disease and decreases atherosclerotic risk in Sprague–Dawley Rats." *Nutrients* 9, no. 7 (2017): 766.

Ali, Abuzer, Mohammad Jameel, and Mohammed Ali. "New fatty acid and acyl glycoside from the aerial parts of Phyllanthus fraternus Webster." *Journal of Pharmacy & Bioallied Sciences* 8, no. 1 (2016): 43–46.

Alvari, Amene, Mehrnaz S. Ohadi Rafsanjani, Farhan J. Ahmed, Mohd S. Hejazi, and Malik Z. Abdin. "Rapid RP-HPLC technique for the determination of phyllanthin as bulk and its quantification in Phyllanthus amarus extract." *International Journal of Phytomedicine* 3, no. 1 (2011): 115–119.

Amessis-Ouchemoukh, Nadia, Ibrahim M. Abu-Reidah, Rosa Quirantes-Piné, Celia Rodríguez-Pérez, Khodir Madani, Alberto Fernández-Gutiérrez, and Antonio Segura-Carretero. "Tentative characterisation of iridoids, phenylethanoid glycosides and flavonoid derivatives from Globularia alypum L. (Globulariaceae) leaves by LC-ESI-QTOF-MS." *Phytochemical Analysis* 25, no. 5 (2014): 389–398.

Amin, Zahra A., Mohammed A. Alshawsh, Mustafa Kassim, Hapipah M. Ali, and Mahmood A. Abdulla. "Gene expression profiling reveals underlying molecular mechanism of hepatoprotective effect of Phyllanthus niruri on thioacetamide-induced hepatotoxicity in Sprague Dawley rats." *BMC Complementary and Alternative Medicine* 13, no. 1 (2013): 160.

Amonkan, Augustin K., Mamadou Kamagaté, Alain NR Yao, André B. Konan, Mathieu N. Kouamé, Camille Koffi, Séraphin Kati-Coulibaly, and Henri Die-Kakou. "Comparative effects of two fractions of Phyllanthus amarus (Euphorbiaceae) on the blood pressure in rabbit." *Greener Journal of Medical Sciences* 3, no. 4 (2013): 129–134.

Anila, L., and N. R. Vijayalakshmi. "Flavonoids from Emblica officinalis and Mangifera indica—effectiveness for dyslipidemia." *Journal of Ethnopharmacology* 79, no. 1 (2002): 81–87.

Annamalai, A., and P. T. V. Lakshmi. "HPTLC and HPLC analysis of bioactive phyllanthin from different organs of Phyllanthus amarus." *Asian Journal of Biotechnology* 1 (2009): 154–162.

Arun, T., B. Senthilkumar, K. Purushothaman, and A. Aarthy. "GC-MS determination of bioactive components of Phyllanthus amarus (L.) and its antibacterial activity." *Journal of Pharmacy Research* 5 (2012): 4767–4771.

Bagalkotkar, G., S. R. Sagineedu, M. S. Saad, and J. Stanslas. "Phytochemicals from Phyllanthus niruri Linn. and their pharmacological properties: a review." *Journal of Pharmacy and Pharmacology* 58, no. 12 (2006): 1559–1570.

Bajpai, Vikas, Sunil Kumar, Awantika Singh, Jyotsana Singh, M. P. S. Negi, Sumit Kumar Bag, Nikhil Kumar, Rituraj Konwar, and Brijesh Kumar. "Chemometric based identification and validation of specific chemical markers for geographical, seasonal and gender variations in Tinospora cordifolia stem using HPLC-ESI-QTOF-MS analysis." *Phytochemical Analysis* 28, no. 4 (2017): 277–288.

Balammal, G., Babu M. Sekar, and J. P. Reddy. "Analysis of herbal medicines by modern chromatographic techniques." *International Journal of Preclinical and Pharmaceutical* Research 3, no. 1 (2012): 50–63.

Bandyopadhyay, Sandip K., Satyesh C. Pakrashi, and Anita Pakrashi. "The role of antioxidant activity of Phyllanthus emblica fruits on prevention from indomethacin induced gastric ulcer." *Journal of Ethnopharmacology* 70, no. 2 (2000): 171–176.

Barros, Marcio E., Roberta Lima, Lucildes P. Mercuri, Jivaldo R. Matos, Nestor Schor, and Mirian A. Boim. "Effect of extract of Phyllanthus niruri on crystal deposition in experimental urolithiasis." *Urological Research* 34, no. 6 (2006): 351–357.

Bhandari, Prasan R., and Mohammad Ameeruddin Kamdod. "Emblica officinalis (Amla): a review of potential therapeutic applications." *International Journal of Green Pharmacy (IJGP)* 6, no. 4 (2012): 257–269.

Boim, Mirian A., Ita P. Heilberg, and Nestor Schor. "Phyllanthus niruri as a promising alternative treatment for nephrolithiasis." *International Brazilian Journal of Urology* 36, no. 6 (2010): 657–664.

Calixto, João B., Adair R. S. Santos, Valdir Cechinel Filho, and Rosendo A. Yunes. "A review of the plants of the genus Phyllanthus: their chemistry, pharmacology, and therapeutic potential." *Medicinal Research Reviews* 18, no. 4 (1998): 225–258.

Chandra, Preeti, Renu Pandey, Mukesh Srivastva, and Brijesh Kumar. "Quality control assessment of polyherbal formulation based on a quantitative determination multimarker approach by ultra-high performance liquid chromatography with tandem mass spectrometry using polarity switching combined with multivariate analysis." *Journal of Separation Science* 38, no. 18 (2015): 3183–3191.

Charoenteeraboon, Juree, Chatri Ngamkitidechakul, Noppamas Soonthornchareonnon, Kanjana Jaijoy, and Seewaboon Sireeratawong. "Antioxidant activities of the standardized water extract from fruit of Phyllanthus emblica Linn." *Sonklanakarin Journal of Science and Technology* 32, no. 6 (2010): 599.

Chaudhary, L. B., and R. R. Rao. "Taxonomic study of herbaceous species of Phyllanthus L. Euphorbiaceae) in India." *Phytotaxonomy* 2 (2002): 143–162.

Chauhan, Rajendra S., Mohan C. Nautiyal, Roberto Cecotti, Mariella Mella, and Aldo Tava. "Variation in the essential oil composition of Angelica archangelica from three different altitudes in Western Himalaya, India." *Industrial Crops and Products* 94 (2016): 401–404.

Chen, Qinhua, Yulin Zhang, Wenpeng Zhang, and Zilin Chen. "Identification and quantification of oleanolic acid and ursolic acid in Chinese herbs by liquid chromatography–ion trap mass spectrometry." *Biomedical Chromatography* 25, no. 12 (2011): 1381–1388.

Chernetsova, Elena S., Elizabeth A. Crawford, Alexander N. Shikov, Olga N. Pozharitskaya, Valery G. Makarov, and Gertrud E. Morlock. "ID-CUBE direct analysis in real time high-resolution mass spectrometry and its capabilities in the identification of phenolic components from the green leaves of Bergenia crassifolia L." *Rapid Communications in Mass Spectrometry* 26, no. 11 (2012): 1329–1337.

Cody, Robert B., James A. Laramée, and H. Dupont Durst. "Versatile new ion source for the analysis of materials in open air under ambient conditions." *Analytical Chemistry* 77, no. 8 (2005): 2297–2302.

Colombo, Renata, Andrea N. de L. Batista, Helder L. Teles, Geraldo H. Silva, Giovani C. C. Bomfim, Rosilene C. R. Burgos, Alberto J. Cavalheiro, et al. "Validated HPLC method for the standardization of Phyllanthus niruri (herb and commercial extracts) using corilagin as a phytochemical marker." *Biomedical Chromatography* 23, no. 6 (2009): 573–580.

Corciovă, Andreia, Cornelia Mircea, Cristina Tuchiluş, Oana Cioancă, Ana-Flavia Burlec, Bianca Ivănescu, Laurian Vlase, et al. "Phenolic and sterolic profile of a phyllanthus amarus extract and characterization of newly synthesized silver nanoparticles." *Farmacia* 66, no. 5 (2018): 831–838.

Cuyckens, Filip, and Magda Claeys. "Mass spectrometry in the structural analysis of flavonoids." *Journal of Mass Spectrometry* 39, no. 1 (2004): 1–15.

da Fontoura Sprenger, Ricardo, and Quezia Bezerra Cass. "Characterization of four Phyllanthus species using liquid chromatography coupled to tandem mass spectrometry." *Journal of Chromatography A* 1291 (2013): 97–103.

Dasaroju, Swetha, and Krishna Mohan Gottumukkala. "Current trends in the research of Emblica officinalis (Amla): a pharmacological perspective." *International Journal of Pharmaceutical Sciences Review and Research* 24, no. 2 (2014): 150–159.

Deepak, P., and G. V. Gopal. "GC-MS analysis of ethyl acetate extract of Phyllanthus emblica L. bark." *British Biomedical Bulletin* 2, no. 2 (2014): 285–292.

Dhiman, Radha K., and Yogesh K. Chawla. "Herbal medicines for liver diseases." *Digestive Diseases and Sciences* 50, no. 10 (2005): 1807–1812.

Dincheva, I., I. Badjakov, V. Kondakova, P. Dobson, G. Mcdougall, and D. Stewart. "Identification of the phenolic components in Bulgarian raspberry cultivars by LC-ESI-MSn." *International Journal of Agricultural Science and Research (IJASR)* 3, no. 3 (2013): 127–138.

Douglas, J. A., M. H. Douglas, D. R. Lauren, R. J. Martin, B. Deo, J. M. Follett, and D. J. Jensen. "Effect of plant density and depth of harvest on the production and quality of licorice (Glycyrrhiza glabra) root harvested over 3 years." *New Zealand Journal of Crop and Horticultural Science* 32, no. 4 (2004): 363–373.

D'souza, Jason Jerome, Prema Pancy D'souza, Farhan Fazal, Ashish Kumar, Harshith P. Bhat, and Manjeshwar Shrinath Baliga. "Anti-diabetic effects of the Indian indigenous fruit Emblica officinalis Gaertn: active constituents and modes of action." *Food & Function* 5, no. 4 (2014): 635–644.

Dutra, Richard Pereira, Bruno Vinicius de Barros Abreu, Mayara Soares Cunha, Marisa Cristina Aranha Batista, Luce Maria Brandão Torres, Flavia Raquel Fernandes Nascimento, Maria Nilce Sousa Ribeiro, and Rosane Nassar Meireles Guerra. "Phenolic acids, hydrolyzable tannins, and antioxidant activity of geopropolis from the stingless bee Melipona fasciculata Smith." *Journal of Agricultural and Food Chemistry* 62, no. 12 (2014): 2549–2557.

Eklund, Patrik C., M. Josefin Backman, Leif Å. Kronberg, Annika I. Smeds, and Rainer E. Sjöholm. "Identification of lignans by liquid chromatography-electrospray ionization ion-trap mass spectrometry." *Journal of Mass Spectrometry* 43, no. 1 (2008): 97–107.

El Amir, Dalia, Sameh F. AbouZid, Mona H. Hetta, Abdelaaty A. Shahat, and Mohamed A. El-Shanawany. "Composition of the essential oil of the fruits of Phyllanthus emblica cultivated in Egypt." *Journal of Pharmaceutical, Chemical and Biological Sciences (JPCBS)* 2, no. 3 (2014): 202–207.

Fabre, Nicolas, Isabelle Rustan, Edmond de Hoffmann, and Joëlle Quetin-Leclercq. "Determination of flavone, flavonol, and flavanone aglycones by negative ion liquid chromatography electrospray ion trap mass spectrometry." *Journal of the American Society for Mass Spectrometry* 12, no. 6 (2001): 707–715.

Fan, Hongyan, Wei Zhang, Jing Wang, Mengying Lv, Pei Zhang, Zunjian Zhang, and Fengguo Xu. "HPLC–MS/MS method for the determination of four lignans from Phyllanthus urinaria L. in rat plasma and its application." *Bioanalysis* 7, no. 6 (2015): 701–712.

Figueiredo, A. Cristina, José G. Barroso, Luis G. Pedro, and Johannes J. C. Scheffer. "Factors affecting secondary metabolite production in plants: volatile components and essential oils." *Flavour and Fragrance Journal* 23, no. 4 (2008): 213–226.

Foo, L. Yeap. "Amarulone, a novel cyclic hydrolysable tannin from Phyllanthus amarus." *Natural Product Letters* 3, no. 1 (1993): 45–52.

Foo, L. Yeap. "Amariinic acid and related ellagitannins from Phyllanthus amarus." *Phytochemistry* 39, no. 1 (1995): 217–224.

Foo, L. Yeap, and Herbert Wong. "Phyllanthusiin D, an unusual hydrolysable tannin from Phyllanthus amarus." *Phytochemistry* 31, no. 2 (1992): 711–713.

Freitas, A. M., Nestor Schor, and Mirian Aparecida Boim. "The effect of Phyllanthus niruri on urinary inhibitors of calcium oxalate crystallization and other factors associated with renal stone formation." *BJU International* 89, no. 9 (2002): 829–834.

Gad, Haidy A., Sherweit H. El-Ahmady, Mohamed I. Abou-Shoer, and Mohamed M. Al-Azizi. "Application of chemometrics in authentication of herbal medicines: a review." *Phytochemical Analysis* 24, no. 1 (2013): 1–24.

Gaire, Bhakta Prasad, and Lalita Subedi. "Phytochemistry, pharmacology and medicinal properties of Phyllanthus emblica Linn." *Chinese Journal of Integrative Medicine* (2014): 1–8, https://doi.org/10.1007/s11655-014-1984-2.

Ghosal, Sibnath, Tripathi, V. K., and Chauhan, S. "Active constituents of Emblica officinalis: part I. The chemistry and antioxidative effects of two new hydrolysable tannins, Emblicanin A and B." *Indian Journal of Chemistry* 35 (1996): 941–948.

Girach, R. D., Aminuddin, P. A. Siddiqui, and Subhan A. Khan. "Traditional plant remedies among the Kondh of District Dhenkanal (Orissa)." *International Journal of Pharmacognosy* 32, no. 3 (1994): 274–283.

Golechha, Mahaveer, Vikas Sarangal, Shreesh Ojha, Jagriti Bhatia, and Dharmveer S. Arya. "Anti-inflammatory effect of Emblica officinalis in rodent models of acute and chronic inflammation: involvement of possible mechanisms." *International Journal of Inflammation* 2014 (2014), http://dx.doi.org/10.1155/2014/178408.

Goodarzi, Mohammad, Paul J. Russell, and Yvan Vander Heyden. "Similarity analyses of chromatographic herbal fingerprints: a review." *Analytica Chimica Acta* 804 (2013): 16–28.

Guideline, ICH Harmonised Tripartite. "Validation of analytical procedures: text and methodology Q2 (R1)." *International Conference on Harmonization*, Geneva, Switzerland, 2005, pp. 11–12.

Gupta, Jyoti, and Mohd Ali. "Four new seco-sterols of Phyllanthus fraternus roots." *Indian Journal of Pharmaceutical Sciences* 61, no. 2 (1999): 90.

Gupta, Jyoti, and Mohd Ali. "Two new oxygenated heterocyclic constituents from Phyllanthus fraternus roots." *Indian Journal of Pharmaceutical Sciences* 62, no. 6 (2000): 438–440.

Habib-ur-Rehman, Khawaja Ansar Yasin, Muhammad Aziz Choudhary, Naeem Khaliq, Atta-Ur-Rahman, Muhammad Iqbal Choudhary, and Shahid Malik. "Studies on the chemical constituents of Phyllanthus emblica." *Natural Product Research* 21, no. 9 (2007): 775–781.

Hajslova, Jana, Tomas Cajka, and Lukas Vaclavik. "Challenging applications offered by direct analysis in real time (DART) in food-quality and safety analysis." *TrAC Trends in Analytical Chemistry* 30, no. 2 (2011): 204–218.

Hamrapurkar, P. D., and S. B. Pawar. "Quantitative estimation of phyllanthin in Phyllanthus amarus using high performance liquid chromatography." *Indian Drugs* 46, no. 4 (2009): 358–360.

Han, Weili, Xiangxiang Chen, Huimin Yu, Lingyun Chen, and Mei Shen. "Seasonal variations of iminosugars in mulberry leaves detected by hydrophilic interaction chromatography coupled with tandem mass spectrometry." *Food Chemistry* 251 (2018): 110–114.

Harish, R., and T. Shivanandappa. "Antioxidant activity and hepatoprotective potential of Phyllanthus niruri." *Food Chemistry* 95, no. 2 (2006): 180 185.

Hossain, Mohammad B., Dilip K. Rai, Nigel P. Brunton, Ana B. Martin-Diana, and Catherine Barry-Ryan. "Characterization of phenolic composition in Lamiaceae spices by LC-ESI-MS/MS." *Journal of Agricultural and Food Chemistry* 58, no. 19 (2010): 10576–10581.

Houghton, Peter J., Tibebe Z. Woldemariam, Siobhan O'Shea, and S. P. Thyagarajan. "Two securinega-type alkaloids from Phyllanthus amarus." *Phytochemistry* 43, no. 3 (1996): 715–717.

Hu, Xueyan, Yunbing Shen, Shengnan Yang, Wei Lei, Cheng Luo, Yuanyuan Hou, and Gang Bai. "Metabolite identification of ursolic acid in mouse plasma and urine after oral administration by ultra-high performance liquid chromatography/quadrupole time-of-flight mass spectrometry." *RSC Advances* 8, no. 12 (2018): 6532–6539.

Huang, Ray-Ling, Yu-Ling Huang, Jun-Chih Ou, Chien-Chih Chen, Feng-Lin Hsu, and Chungming Chang. "Screening of 25 compounds isolated from Phyllanthus species for anti-human hepatitis B virus in vitro." *Phytotherapy Research* 17, no. 5 (2003): 449–453.

Hvattum, Erlend, and Dag Ekeberg. "Study of the collision-induced radical cleavage of flavonoid glycosides using negative electrospray ionization tandem quadrupole mass spectrometry." *Journal of Mass Spectrometry* 38, no. 1 (2003): 43–49.

Ibrahim, A. M., B. Lawal, A. N. Abubakar, N. A. Tsado, G. N. Kontagora, J. A. Gboke, and E. B. Berinyuy. "Antimicrobial and Free Radical Scavenging Potentials of N-Hexane and Ethyl Acetate Fractions of Phyllanthus Fraternus." *Nigerian Journal of Basic and Applied Sciences* 25, no. 2 (2017): 6–11.

Igwe, O. U., and F. U. Okwunodulu. "Investigation of bioactive phytochemical compounds from the chloroform extract of the leaves of phyllanthus amarus by GC-MS technique." *International Journal of Chemistry and Pharmaceutical* 2, no. 1 (2014): 554–560.

Ishimaru, Kanji, Kayo Yoshimatsu, Takashi Yamakawa, Hiroshi Kamada, and Koichiro Shimomura. "Phenolic constituents in tissue cultures of Phyllanthus niruri." *Phytochemistry* 31, no. 6 (1992): 2015–2018.

Iswaldi, Ihsan, Ana María Gómez-Caravaca, Jesús Lozano-Sánchez, David Arráez-Román, Antonio Segura-Carretero, and Alberto Fernández-Gutiérrez. "Profiling of phenolic and other polar compounds in zucchini (Cucurbita pepo L.) by reverse-phase high-performance liquid chromatography coupled to quadrupole time-of-flight mass spectrometry." *Food Research International* 50, no. 1 (2013): 77–84.

Jantan, Ibrahim, Menaga Ilangkovan, and Hazni Falina Mohamad. "Correlation between the major components of Phyllanthus amarus and Phyllanthus urinaria and their inhibitory effects on phagocytic activity of human neutrophils." *BMC Complementary and Alternative Medicine* 14, no. 1 (2014): 429.

Jiang, Yong, Bruno David, Pengfei Tu, and Yves Barbin. "Recent analytical approaches in quality control of traditional Chinese medicines—a review." *Analytica Chimica Acta* 657, no. 1 (2010): 9–18.

Jin, Miao-miao, Wen-dan Zhang, Yan-mei Xu, Ying-feng Du, Qian Sun, Wei Guo, Liang Cao, and Hui-jun Xu. "Simultaneous determination of 12 active components in the roots of Pulsatilla chinensis using tissue-smashing extraction with liquid chromatography and mass spectrometry." *Journal of Separation Science* 40, no. 6 (2017): 1283–1292.

Joshi, Devi Datt. *Herbal Drugs and Fingerprints: Evidence Based Herbal Drugs.* New Delhi: Springer Science & Business Media, 2012.

Joy, K. L., and R. Kuttan. "Anti-oxidant activity of selected plant extract." *Amala Research Bulletin* 15 (1995): 68–71.

Karar, M. G. Elsadig, and N. Kuhnert. "UPLC-ESI-Q-TOF-MS/MS characterization of phenolics from Crataegus monogyna and Crataegus laevigata (Hawthorn) leaves, fruits and their herbal derived drops (Crataegutt Tropfen)." *Journal of Chemical Biology & Therapeutics* 1 (2015): 1–23.

Kassuya, Cândida A. L., Aline Silvestre, Octavio Menezes-de-Lima Jr, Denise Mollica Marotta, Vera Lúcia G. Rehder, and João B. Calixto. "Antiinflammatory and antiallodynic actions of the lignan niranthin isolated from Phyllanthus amarus: evidence for interaction with platelet activating factor receptor." *European Journal of Pharmacology* 546, no. 1–3 (2006): 182–188.

Ketmongkhonsit, Pakabhorn, Chaiyo Chaichantipyuth, Chanida Palanuvej, Worathat Thitikornpong, and Suchada Sukrong. "A validated TLC-image analysis method for detecting and quantifying bioactive phyllanthin in Phyllanthus amarus and commercial herbal drugs." *Songklanakarin Journal of Science & Technology* 37, no. 3 (2015): 319–326.

Khan, Kishwar Hayat. "Roles of Emblica officinalis in medicine-A review." *Botany Research International* 2, no. 4 (2009): 218–228.

Khan, Salim, Fahad Al-Qurainy, Mauji Ram, Sayeed Ahmad, and Malik Zainul Abdin. "Phyllanthin biosynthesis in Phyllanthus amarus: Schum and Thonn growing at different altitudes." *Journal of Medicinal Plants Research* 4, no. 1 (2010): 41–48.

Khan, Salim, Rajeev K. Singla, and Malik Zainul Abdin. "Assessment of phytochemical diversity in Phyllanthus amarus using HPTLC fingerprints." *Indo Global Journal of Pharmaceutical Sciences* 1, no. 1 (2011): 1–12.

Khanna, P., and R. Bansal. "Phyllantidine & phyllantine from Emblica officinalis Gaertn leaves fruits & in vitro tissue cultures." *Indian Journal of Experimental Biology* 13 (1975): 82–83.

Khanna, A. K., F. Rizvi, and R. Chander. "Lipid lowering activity of Phyllanthus niruri in hyperlipemic rats." *Journal of Ethnopharmacology* 82, no. 1 (2002): 19–22.

Kirtikar, K. R., and B. D. Basu. *Medicinal Plants of India* (Vol. 3, 2nd ed., pp. 1559–1560). Allahabad: Lalit Mohan Basu Publications, 1933.

Kirtikar, K. R., and B. D. Basu. *Indian Medicinal Plants* (Vol. 1, 2nd ed.). Allahabad: Lalit Mohan Babu and Co., 1975.

Koffuor, G. A., and P. Amoateng. "Antioxidant and anticoagulant properties of Phyllanthus fraternus GL Webster (Family: Euphorbiaceae)." *Journal of Pharmacology and Toxicology* 6, no. 7 (2011): 624–636.

Komlaga, Gustav, Sandrine Cojean, Rita A. Dickson, Mehdi A. Beniddir, Soulaf Suyyagh-Albouz, Merlin L. K. Mensah, Christian Agyare, Pierre Champy, and Philippe M. Loiseau. "Antiplasmodial activity of selected medicinal plants used to treat malaria in Ghana." *Parasitology Research* 115, no. 8 (2016): 3185–3195.

Komlaga, Gustav, Grégory Genta Jouvo, Sandrino Cojean, Rita A. Dickson, Merlin L. K. Mensah, Philippe M. Loiseau, Pierre Champy, and Mehdi A. Beniddir. "Antiplasmodial Securinega alkaloids from Phyllanthus fraternus: discovery of natural (+)-allonorsecurinine." *Tetrahedron Letters* 58, no. 38 (2017): 3754 3756.

Kumar, K. B. H., and R. Kuttan. "Chemoprotective activity of an extract of Phyllanthus amarus against cyclophosphamide induced toxicity in mice." *Phytomedicine* 12, no. 6–7 (2005): 494–500.

Kumar, Sandeep, Amita Yadav, Manila Yadav, and Jaya Parkash Yadav. "Effect of climate change on phytochemical diversity, total phenolic content and in vitro antioxidant activity of Aloe vera (L.) Burm. f." *BMC Research Notes* 10, no. 1 (2017a): 60.

Kumar, Sunil, Awantika Singh, and Brijesh Kumar. "Identification and characterization of phenolics and terpenoids from ethanolic extracts of Phyllanthus species by HPLC-ESI-QTOF-MS/MS." *Journal of Pharmaceutical Analysis* 7, no. 4 (2017b): 214–222.

Kumar, Sunil, Awantika Singh, Vikas Bajpai, Bikarma Singh, and Brijesh Kumar. "Development of a UHPLC–MS/MS method for the quantitation of bioactive compounds in Phyllanthus species and its herbal formulations." *Journal of Separation Science* 40, no. 17 (2017c): 3422–3429.

Kumar, Sunil, Preeti Chandra, Vikas Bajpai, Awantika Singh, Mukesh Srivastava, D. K. Mishra, and Brijesh Kumar. "Rapid qualitative and quantitative analysis of bioactive compounds from Phyllanthus amarus using LC/MS/MS techniques." *Industrial Crops and Products* 69 (2015a): 143–152.

Kumar, Sunil, Vikas Bajpai, Awantika Singh, S. Bindu, Mukesh Srivastava, K. B. Rameshkumar, and Brijesh Kumar. "Rapid fingerprinting of Rauwolfia species using direct analysis in real time mass spectrometry combined with principal component analysis for their discrimination." *Analytical Methods* 7, no. 14 (2015b): 6021–6026.

Kumar, Sunil, Vikas Bajpai, Mukesh Srivastava, and Brijesh Kumar. "Study of geographical variation in Phyllanthus amarus Schum & Thonn using DART-TOF-MS combined with PCA." *The Brazilian Journal of Analytical Chemistry (BrJAC)* 4, no. 17 (2017d): 16–23.

Kumaran, A., and R. Joel Karunakaran. "In vitro antioxidant activities of methanol extracts of five Phyllanthus species from India." *LWT-Food Science and Technology* 40, no. 2 (2007): 344–352.

Kumaran, A., and R. Joel Karunakaran. "Nitric oxide radical scavenging active components from Phyllanthus emblica L." *Plant Foods for Human Nutrition* 61, no. 1 (2006): 1–5.

Kuo, Ching-Hua, Shoei-Sheng Lee, Hsing-Yun Chang, and Shao-Wen Sun. "Analysis of lignans using micellar electrokinetic chromatography." *Electrophoresis* 24, no. 6 (2003): 1047–1053.

Kuttan, Ramadasan, and K. B. Harikumar. *Phyllanthus Species: Scientific Evaluation and Medicinal Applications*. Boca Raton, FL: CRC Press, 2011.

Lee, Kuo-Hsiung, and Zhiyan Xiao. "Lignans in treatment of cancer and other diseases." *Phytochemistry Reviews* 2, no. 3 (2003): 341–362.

Lee, Nathanael YS, William KS Khoo, Mohammad Akmal Adnan, Tanes Prasat Mahalingam, Anne R. Fernandez, and Kamalan Jeevaratnam. "The pharmacological potential of Phyllanthus niruri." *Journal of pharmacy and pharmacology* 68, no. 8 (2016): 953–969.

Liang, Yi-Zeng, Peishan Xie, and Kelvin Chan. "Quality control of herbal medicines." *Journal of Chromatography B* 812, no. 1–2 (2004): 53–70.

Lim, Yau Yan, and Johannes Murtijaya. "Antioxidant properties of Phyllanthus amarus extracts as affected by different drying methods." *LWT-Food Science and Technology* 40, no. 9 (2007): 1664–1669.

Liu, Jian-ping, Hui Lin, and H. McIntosh. "Genus Phyllanthus for chronic hepatitis B virus infection: a systematic review." *Journal of Viral Hepatitis* 8, no. 5 (2001): 358–366.

Liu, X., M. Zhao, K. Wu, X. Chai, H. Yu, Z. Tao, and J. Wang. Immunomodulatory and anticancer activities of phenolics from emblica fruit (Phyllanthus emblica L.). *Food Chemistry* 131, no. 2 (2012): 685–690.

Liu, Xiaoli, Chun Cui, Mouming Zhao, Jinshui Wang, Wei Luo, Bao Yang, and Yueming Jiang. "Identification of phenolics in the fruit of emblica (Phyllanthus emblica L.) and their antioxidant activities." *Food Chemistry* 109, no. 4 (2008): 909–915.

Llorach, Rafael, Angel Gil-Izquierdo, Federico Ferreres, and Francisco A. Tomás-Barberán. "HPLC-DAD-MS/MS ESI characterization of unusual highly glycosylated acylated flavonoids from cauliflower (Brassica oleracea L. v ar. botrytis) agroindustrial byproducts." *Journal of Agricultural and Food Chemistry* 51, no. 13 (2003): 3895–3899.

Londhe, Jayant S., Thomas P. A. Devasagayam, L. Yeap Foo, Padma Shastry, and Saroj S. Ghaskadbi. "Geraniin and amariin, ellagitannins from Phyllanthus amarus, protect liver cells against ethanol induced cytotoxicity." *Fitoterapia* 83, no. 8 (2012): 1562–1568.

Londhe, Jayant S., Thomas P. A. Devasagayam, L. Yeap Foo, and Saroj S. Ghaskadbi. "Radioprotective properties of polyphenols from Phyllanthus amarus Linn." *Journal of Radiation Research* 50, no. 4 (2009): 303–309.

López, Michelle, and Bernd Hoppe. "History, epidemiology and regional diversities of urolithiasis." *Pediatric Nephrology* 25, no. 1 (2010): 49.

Luo, Wei, Mouming Zhao, Bao Yang, Guanglin Shen, and Guohua Rao. "Identification of bioactive compounds in Phyllenthus emblica L. fruit and their free radical scavenging activities." *Food Chemistry* 114, no. 2 (2009): 499–504.

Lv, Jun-Jiang, Ya-Feng Wang, Jing-Min Zhang, Shan Yu, Dong Wang, Hong-Tao Zhu, Rong-Rong Cheng, Chong-Ren Yang, Min Xu, and Ying-Jun Zhang. "Anti-hepatitis B virus activities and absolute configurations of sesquiterpenoid glycosides from Phyllanthus emblica." *Organic & Biomolecular Chemistry* 12, no. 43 (2014): 8764–8774.

MacRae, W. Donald, and G. H. Neil Towers. "Biological activities of lignans." *Phytochemistry* 23, no. 6 (1984): 1207–1220.

Mahdi, Elrashid Saleh, Azmin Mohd Noor, Mohamed Hameem Sakeena, Ghassan Z. Abdullah, Muthanna Abdulkarim, and Munavvar Abdul Sattar. "Identification of phenolic compounds and assessment of in vitro antioxidants activity of 30% ethanolic extracts derived from two Phyllanthus species indigenous to Malaysia." *African Journal of Pharmacy and Pharmacology* 5, no. 17 (2011): 1967–1978.

Maity, Soumya, Suchandra Chatterjee, Prasad Shekhar Variyar, Arun Sharma, Soumyakanti Adhikari, and Santasree Mazumder. "Evaluation of antioxidant activity and characterization of phenolic constituents of Phyllanthus amarus root." *Journal of Agricultural and Food Chemistry* 61, no. 14 (2013): 3443–3450.

Mamza, U. T., O. A. Sodipo, and I. Z. Khan. "Gas chromatography–mass spectrometry (GC–MS) analysis of bioactive components of Phyllanthus amarus leaves." *International Research Journal of Plant Science* 3, no. 10 (2012): 208–215.

Mao, Xin, Ling-Fang Wu, Hong-Ling Guo, Wen-Jing Chen, Ya-Ping Cui, Qi, Shi Li, et al. "The genus Phyllanthus: an ethnopharmacological, phytochemical, and pharmacological review." *Evidence-Based Complementary and Alternative Medicine* 2016 (2016): Article ID: 7584952, 36 pages.

Martins, Lucia Regina Rocha, Edenir Rodrigues Pereira-Filho, and Quezia Bezerra Cass. "Chromatographic profiles of Phyllanthus aqueous extracts samples: a proposition of classification using chemometric models." *Analytical and Bioanalytical Chemistry* 400, no. 2 (2011): 469–481.

Matur, B. M., T. Matthew, and C. I. C. Ifeanyi. "Analysis of the phytochemical and in vivo antimalaria properties of Phyllanthus fraternus Webster extract." *New York Science Journal* 2, no. 5 (2009): 12–19.

Mazumder, Avijit, Arun Mahato, and Rupa Mazumder. "Antimicrobial potentiality of Phyllanthus amarus against drug resistant pathogens." *Natural Product Research* 20, no. 4 (2006): 323–326.

Mediani, Ahmed, Faridah Abas, M. Maulidiani, Alfi Khatib, Chin Ping Tan, Intan Safinar Ismail, Khozirah Shaari, and Amin Ismail. "Characterization of Metabolite Profile in Phyllanthus niruri and Correlation with Bioactivity Elucidated by Nuclear Magnetic Resonance Based Metabolomics." *Molecules* 22, no. 6 (2017): 902.

Mehta, Kavit, Patel, B. N., and Jain, B. K. "Phytochemical analysis of leaf extract of Phyllanthus fraternus." *Research Journal of Recent Sciences* 2 (2013): 12–15.

Moirangthem, R. S., S. V. J. Singh, H. C. Devi, K. H. Devi, N. Gunindro, and M. N. Devi. "Effect of aqueous extract of Phyllanthus fraternus leaf against cyclophosphamide induced dyslipidemia and aortitis in wistar albino rats." *International Journal of Contemporary Medical Research* 3, no. 9 (2016): 2703–2706.

Mojarrab, Mahdi, Marjan-Sadat Lagzian, Seyed Ahmad Emami, Javad Asili, and Zahra Tayarani-Najaran. "In vitro anti-proliferative and apoptotic activity of different fractions of Artemisia armeniaca." *Revista Brasileira de Farmacognosia* 23, no. 5 (2013): 783–788.

Murali, B., A. Amit, M. S. Anand, T. K. Dinesh, and D. S. Samiulla. "An improved HPLC method for estimation of phyllanthin and hypophyllanthin in Phyllanthus amarus." *Journal of Natural Remedies* 1, no. 1 (2001): 55–59.

Murugaiyah, Vikneswaran, and Kit Lam Chan. "Analysis of lignans from Phyllanthus niruri L. in plasma using a simple HPLC method with fluorescence detection and its application in a pharmacokinetic study." *Journal of Chromatography B* 852, no. 1–2 (2007): 138–144.

Muthusamy, A., E. R. Sanjay, HN Nagendra Prasad, M. Radhakrishna Rao, B. Manjunath Joshi, S. Padmalatha Rai, and K. Satyamoorthy. "Quantitative analysis of phyllanthus species for bioactive molecules using high-pressure liquid chromatography and liquid chromatography–mass spectrometry." *Proceedings of the National Academy of Sciences, India Section B: Biological Sciences* 88, no. 3 (2018): 1043–1054.

Nadkarni, K. M., and A. K. Nadkarni. Indian Materia Medica, Popular Prakashan Pvt. Ltd., Bombay 1 (1976): 799.

Narendra, K., J. Swathi, K. M. Sowjanya, and A. Krishna Satya. "Phyllanthus niruri: a review on its ethno botanical, phytochemical and pharmacological profile." *Journal of Pharmacy Research* 5, no. 9 (2012): 4681–4691.

Nasrulloh, Roni, Mohamad Rafi, Wulan Tri Wahyuni, Shuichi Shimma, and Rudi Heryanto. "HPLC fingerprint and simultaneous quantitative analysis of phyllanthin and hypophyllanthin for identification and authentication of Phyllanthus niruri from related species." *Revista Brasileira de Farmacognosia* 28, no. 5 (2018): 527–532.

Natural Medicines. Natural Standard Professional Monograph. Chanca Piedra. "http://naturaldatabase.therapeuticresearch.com/nd/ (Accessed December 26 2018).

Navarro, Mirtha, Ileana Moreira, Elizabeth Arnaez, Silvia Quesada, Gabriela Azofeifa, Felipe Vargas, Diego Alvarado, and Pei Chen. "Flavonoids and ellagitannins characterization, antioxidant and cytotoxic activities of Phyllanthus acuminatus Vahl." *Plants* 6, no. 4 (2017): 62.

Nayak, Preeti Sagar, Anubha Upadhyay, Sunil Kumar Dwivedi, and R. A. O. Sathrupa. "Quantitative determination of phyllanthin in Phyllanthus amarus by high-performance thin layer chromatography." *Boletín Latinoamericano y del Caribe de Plantas Medicinales y Aromáticas* 9, no. 5 (2010): 353–358.

Nisar, Muhammad Farrukh, Junwei He, Arsalan Ahmed, Youxin Yang, Mingxi Li, and Chunpeng Wan. "Chemical components and biological activities of the genus Phyllanthus: A review of the recent literature." *Molecules* 23, no. 10 (2018): 2567.

Niu, Xiaofeng, Lin Qi, Weifeng Li, and Xia Liu. "Simultaneous analysis of eight phenolic compounds in Phyllanthus simplex Retz by HPLC-DAD-ESI/MS." *Journal of Medicinal Plants Research* 6, no. 9 (2012): 1512–1518.

Notka, Frank, Georg Meier, and Ralf Wagner. "Concerted inhibitory activities of Phyllanthus amarus on HIV replication in vitro and ex vivo." *Antiviral Research* 64, no. 2 (2004): 93–102.

Ofuegbe, Sunday Oluwaseun, Adeolu Alex Adedapo, and Abiodun Adegoke Adeyemi. "Anti-inflammatory and analgesic activities of the methanol leaf extract of Phyllanthus amarus in some laboratory animals." *Journal of Basic and Clinical Physiology and Pharmacology* 25, no. 2 (2014): 175–180.

Oseni, L. A., D. Amiteye, S. Antwi, M. Tandoh, and G. M. Aryitey. "Preliminary in vivo evaluation of anti-inflammatory activities of aqueous and ethanolic whole plant extracts of *Phyllantus fraternus* on Carrageenan-induced Paw Oedema in Sprague-Dawley Rats." *Journal of Applied Pharmaceutical Science* 3, no. 3 (2013): 62–65.

Ott, M., Thyagarajan, S. P., and Gupta, S. Phyllanthus amarus suppresses hepatitis B virus by interrupting interactions between HBV enhancer I and cellular transcription factors. *European Journal of Clinical Investigation* 27 (1997): 908–915.

Packirisamy, Rajaa Muthu, Zachariah Bobby, Sankar Panneerselvam, Smitha Mariam Koshy, and Sajini Elizabeth Jacob. "Metabolomic analysis and antioxidant effect of Amla (Emblica officinalis) extract in preventing oxidative stress-induced red cell damage and plasma protein alterations: an in vitro study." *Journal of Medicinal Food* 21, no. 1 (2018): 81–89.

Paranjape, P. *Indian Medicinal Plants. Forgotten Healer: A Guide to Ayurvedic Herbal Medicine* (pp. 148–149). Delhi: Chaukhamba Sanskrit Pratisthan, 2001.

Patel, J. R., P. Tripathi, V. Sharma, N. S. Chauhan, and V. K. Dixit. Phyllanthus amarus: ethnomedicinal uses, phytochemistry and pharmacology: a review. *Journal of Ethnopharmacology* 138, no. 2 (2011): 286–313.

Patel, S. S., and R. K. Goyal. "Emblica officinalis Geart.: a comprehensive review on phytochemistry, pharmacology and ethnomedicinal uses." *Research Journal of Medicinal Plant* 6 (2012): 6–16.

Peñalvo, José L., Kati M. Haajanen, Nigel Botting, and Herman Adlercreutz. "Quantification of lignans in food using isotope dilution gas chromatography/mass spectrometry." *Journal of Agricultural and Food Chemistry* 53, no. 24 (2005): 9342–9347.

Pfundstein, Beate, Samy K. El Desouky, William E. Hull, Roswitha Haubner, Gerhard Erben, and Robert W. Owen. "Polyphenolic compounds in the fruits of Egyptian medicinal plants (Terminalia bellerica, Terminalia chebula and Terminalia horrida): characterization, quantitation and determination of antioxidant capacities." *Phytochemistry* 71, no. 10 (2010): 1132–1148.

Poh-Hwa, T., C. Yoke-Kqueen, J. Indu Bala, and R. Son. "Bioprotective properties of three Malaysia Phyllanthus species: an investigation of the antioxidant and antimicrobial activities." *International Food Research Journal* 18, no. 3 (2011): 887–893.

Pramyothin, Pornpen, Chanon Ngamtin, Somlak Poungshompoo, and Chaiyo Chaichantipyuth. "Hepatoprotective activity of Phyllanthus amarus Schum. et. Thonn. extract in ethanol treated rats: in vitro and in vivo studies." *Journal of Ethnopharmacology* 114, no. 2 (2007): 169–173.

Pucci, Nidia D., Giovanni S. Marchini, Eduardo Mazzucchi, Sabrina T. Reis, Miguel Srougi, Denise Evazian, and William C. Nahas. "Effect of phyllanthus niruri on metabolic parameters of patients with kidney stone: a perspective for disease prevention." *International Brazilian Journal of Urology* 44, no. 4 (2018): 758–764.

Qi, Weiyan, Lei Hua, and Kun Gao. "Chemical constituents of the plants from the genus Phyllanthus." *Chemistry & Biodiversity* 11, no. 3 (2014): 364–395.

Rai, Pallavi, Purushottam Patil, and S. J. Rajput. "Simultaneous determination of phyllanthin and hypophyllanthin in herbal formulation by derivative spectrophotometry and liquid chromatography." *Pharmacognosy Magazine* 5, no. 18 (2009): 151.

Rajeshkumar, N. V., K. L. Joy, Girija Kuttan, R. S. Ramsewak, Muraleedharan G. Nair, and Ramadasan Kuttan. "Antitumour and anticarcinogenic activity of Phyllanthusamarus extract." *Journal of Ethnopharmacology* 81, no. 1 (2002): 17–22.

Rani, Phulan, and Neeraj Khullar. "Antimicrobial evaluation of some medicinal plants for their anti-enteric potential against multi-drug resistant Salmonella typhi." *Phytotherapy Research: An International Journal Devoted to Pharmacological and Toxicological Evaluation of Natural Product Derivatives* 18, no. 8 (2004): 670–673.

Rastogi, Ram P., and B. N. Mehrotra. *Compendium of Indian Medicinal Plants.* Vol. II. New Delhi: Central Drug Research Institute, Lucknow and publications and Information Directorate., 1990.

Sánchez-Rabaneda, Ferran, Olga Jáuregui, Isidre Casals, Cristina Andrés-Lacueva, Maria Izquierdo-Pulido, and Rosa M. Lamuela-Raventós. "Liquid chromatographic/electrospray ionization tandem mass spectrometric study of the phenolic composition of cocoa (Theobroma cacao)." *Journal of Mass Spectrometry* 38, no. 1 (2003a): 35–42.

Sánchez-Rabaneda, Ferran, Olga Jauregui, Rosa Maria Lamuela-Raventos, Jaume Bastida, Francesc Viladomat, and Carles Codina. "Identification of phenolic compounds in artichoke waste by high-performance liquid chromatography–tandem mass spectrometry." *Journal of Chromatography A* 1008, no. 1 (2003b): 57–72.

Santoshkumar, Jeevangi, S. Manjunath, and Pranavkumar M. Sakhare. "A study of anti-hyperlipidemia, hypolipedimic and anti-atherogenic activity of fruit of Emblica officinalis (amla) in high fat fed albino rats." *International Journal of Medical Research & Health Sciences* 1, no. 2 (2013): 70–77.

Santos, Sónia AO, Carmen SR Freire, M. Rosário M. Domingues, Armando J. D. Silvestre, and Carlos Pascoal Neto. "Characterization of phenolic components in polar extracts of Eucalyptus globulus Labill. bark by high-performance liquid chromatography–mass spectrometry." *Journal of Agricultural and Food Chemistry* 59, no. 17 (2011): 9386–9393.

Sawant, Laxman, B. Prabhakar, and N. Pandita. "Quantitative HPLC analysis of ascorbic acid and gallic acid in Phyllanthus emblica." *Journal of Analytical and Bioanalytical Techniques* 1, no. 3 (2010): 111 doi:10.4172/2155-9872.1000111.

Schröder, Heinz C., Helmut Merz, Renate Steffen, Werner E Muller, Prem S. Sarin, Susanne Trumm, Jutta Schulz, and Eckart Eich. "Differential in vitro anti-HIV activity of natural lignans." *Zeitschrift für Naturforschung C* 45, no. 11–12 (1990): 1215–1221.

Shah, Flora C., and Nayan K. Jain. "In Vitro Study on Hepatoprotective effect of Phyllanthus fraternus against lead induced toxicity." *UK Journal of Pharmaceutical and Biosciences* 4, no. 2 (2016): 31–37.

Shah, R. A., S. Khan, P. D. Sonawane, and W. Rehman. "Phytochemical finger printing and antimicrobial activity of Phyllanthus niruri." *International Journal of Pharmaceutical Sciences Review and Research* 44, no. 2 (2017): 7–11.

Shakil, N. A., J. Kumar, R. K. Pandey, and D. B. Saxena. "Nematicidal prenylated flavanones from Phyllanthus niruri." *Phytochemistry* 69, no. 3 (2008): 759–764.

Shanker, K., M. Singh, V. Srivastava, R. K. Verma, A. K. Gupta, and M. M. Gupta. "Simultaneous analysis of six bioactive lignans in Phyllanthus species by reversed phase hyphenated high performance liquid chromatographic technique." *Acta Chromatographica* 23, no. 2 (2011): 321–337.

Sharma, Anupam, Ravneet T. Singh, and Sukhdev S. Handa. "Estimation of phyllanthin and hypophyllanthin by high performance liquid chromatography in Phyllanthus amarus." *Phytochemical Analysis* 4, no. 5 (1993): 226–229.

Shead, Andrew, Karen Vickery, Aniko Pajkos, Robert Medhurst, John Freiman, Robert Dixon, and Yvonne Cossart. "Effects of Phyllanthus plant extracts on duck hepatitis B virus in vitro and in vivo." *Antiviral Research* 18, no. 2 (1992): 127–138.

Shen, Yao, Wan-Ying Wu, and De-An Guo. "DART-MS: a new research tool for herbal medicine analysis." *World Journal of Traditional Chinese Medicine* 2 (2016): 2–9.

Shirani, Majid, Roya Raeisi, Saeid Heidari-Soureshjani, Majid Asadi-Samani, and Tahra Luther. "A review for discovering hepatoprotective herbal drugs with least side effects on kidney." *Journal of Nephropharmacology* 6, no. 2 (2017), 38–48.

Singh, Awantika, Vikas Bajpai, Mukesh Srivastava, Kamal Ram Arya, and Brijesh Kumar. "Rapid screening and distribution of bioactive compounds in different parts of Berberis petiolaris using direct analysis in real time mass spectrometry." *Journal of Pharmaceutical Analysis* 5, no. 5 (2015): 332–335.

Singh, Binnu. "Pharmacological studies on novel anti-diabetic bioactive constituents of some ethno medicinal plants of Mizoram." PhD diss., Mizoram University, 2016.

Singh, Ekta, Sheel Sharma, Ashutosh Pareek, Jaya Dwivedi, Sachdev Yadav, and Swapnil Sharma. "Phytochemistry, traditional uses and cancer chemopreventive activity of Amla (Phyllanthus emblica): the sustainer." *Journal of Applied Pharmaceutical Science* 2, no. 1 (2011): 176–183.

Singh, Khushboo, Manju Panghal, Sangeeta Kadyan, Uma Chaudhary, and Jaya Parkash Yadav. "Green silver nanoparticles of Phyllanthus amarus: as an antibacterial agent against multi drug resistant clinical isolates of Pseudomonas aeruginosa." *Journal of Nanobiotechnology* 12, no. 1 (2014): 40.

Sittie, A. A., E. Lemmich, C. E. Olsen, L. Hviid, and S. Brøgger Christensen. "Alkamides from Phyllanthus fraternus." *Planta Medica* 64, no. 2 (1998): 192–193.

Srirama, Ramanujam, JU Santhosh Kumar, G. S. Seethapathy, Steven G. Newmaster, S. Ragupathy, K. N. Ganeshaiah, R. Uma Shaanker, and Gudasalamani Ravikanth. "Species adulteration in the herbal trade: causes, consequences and mitigation." *Drug Safety* 40, no. 8 (2017): 651–661.

Srivastava, Vandana, Manju Singh, Richa Malasoni, Karuna Shanker, Ram K. Verma, Madan M. Gupta, Anil K. Gupta, and Suman P. S. Khanuja. "Separation and quantification of lignans in Phyllanthus species by a simple chiral densitometric method." *Journal of Separation Science* 31, no. 1 (2008): 47–55.

Sudjaroen, Yuttana, William E. Hull, Gerhard Erben, Gerd Würtele, Supranee Changbumrung, Cornelia M. Ulrich, and Robert W. Owen. "Isolation and characterization of ellagitannins as the major polyphenolic components of Longan (Dimocarpus longan Lour) seeds." *Phytochemistry* 77 (2012): 226–237.

Syamasundar, Kodakandla Venkata, Bikram Singh, Raghunath Singh Thakur, Akhtar Husain, Kiso Yoshinobu, and Hikino Hiroshi. "Antihepatotoxic principles of Phyllanthus niruri herbs." *Journal of Ethnopharmacology* 14, no. 1 (1985): 41–44.

Taylor, L. *Herbal Secrets of the Rainforest* (2nd ed.) New York: Sage Press Incorporated.

Tharakan, S. T. Taxonomy of the genus *Phyllanthus*. In: R. Kuttan, K. Harikumar (eds.), *Phyllanthus Species: Scientific Evaluation and Medicinal Applications* (pp. 23–36). Florida: CRC Press, 2011.

Thilakchand, Karadka Ramdas, Rashmi Teresa Mathai, Paul Simon, Rithin T. Ravi, Manjeshwar Poonam Baliga-Rao, and Manjeshwar Shrinath Baliga. "Hepatoprotective properties of the Indian gooseberry (Emblica officinalis Gaertn): a review." *Food & Function* 4, no. 10 (2013): 1431–1441.

Thyagarajan, S. P., and S. Jayaram. "Natural history of Phyllanthus amarus in the treatment of hepatitis B." *Indian Journal of Medical Microbiology* 10, no. 2 (1992): 64–80.

Thyagarajan, S. P., S. Jayaram, V. Gopalakrishnan, R. Hari, P. Jeyakumar, and M. S. Sripathi. "Herbal medicines for liver diseases in India." *Journal of Gastroenterology and Hepatology* 17 (2002): S370–S376.

Tona, L., K. Mesia, N. P. Ngimbi, B. Chrimwami, Okond ahoka, K. Cimanga, T. de Bruyne, S. Apers, N. Hermans, J. Totte, L. Pieters, and A. J. Vlietinck. In-vivo antimalarial activity of Cassia occidentalis, Morinda morindoides and Phyllanthus niruri. *Annals of Tropical Medicine and Parasitology* 95, no. 1(2001): 47–57.

Tona, L., N. P. Ngimbi, M. Tsakala, K. Mesia, K. Cimanga, S. Apers, T. De Bruyne, L. Pieters, J. Totte, and A. J. Vlietinck. "Antimalarial activity of 20 crude extracts from nine African medicinal plants used in Kinshasa, Congo." *Journal of Ethnopharmacology* 68, no. 1–3 (1999): 193–203.

Treutter, Dieter. "Significance of flavonoids in plant resistance: a review." *Environmental Chemistry Letters* 4, no. 3 (2006): 147.

Tripathi, Arvind K., Ram K. Verma, Anil K. Gupta, Madan M. Gupta, and Suman P. S. Khanuja. "Quantitative determination of phyllanthin and hypophyllanthin in Phyllanthus species by high-performance thin layer chromatography." *Phytochemical Analysis: An International Journal of Plant Chemical and Biochemical Techniques* 17, no. 6 (2006): 394–397.

Tuominen, Anu, and Terhi Sundman. "Stability and oxidation products of hydrolysable tannins in basic conditions detected by HPLC/DAD–ESI/QTOF/MS." *Phytochemical Analysis* 24, no. 5 (2013): 424–435.

Upadhyay, Richa, Jitendra Kumar Chaurasia, Kavindra Nath Tiwari, and Karuna Singh. "Antioxidant property of aerial parts and root of Phyllanthus fraternus Webster, an important medicinal plant." *The Scientific World Journal* 2014 (2014) Article ID: 692392, 7 pages.

Vallverdú-Queralt, Anna, Olga Jáuregui, Alexander Medina-Remón, and Rosa Maria Lamuela-Raventós. "Evaluation of a method to characterize the phenolic profile of organic and conventional tomatoes." *Journal of Agricultural and Food Chemistry* 60, no. 13 (2012): 3373–3380.

Ved, D. K., and G. S. Goraya. *Demand and Supply of Medicinal Plants in India*. Bishen Singh Mahendra Pal Singh. Bangalore, India: FRLHT, 2008.

Verma, Sonia, Hitender Sharma, and Munish Garg. "Phyllanthus amarus: a review." *Journal of pharmacognosy and Phytochemistry* 3, no. 2 (2014): 18–22.

Wang, Cheng-cheng, Jia-rui Yuan, Chun-fei Wang, Nan Yang, Juan Chen, Dan Liu, Jie Song, Liang Feng, Xiao-bin Tan, and Xiao-bin Jia. "Anti-inflammatory effects of Phyllanthus emblica L on benzopyrene-induced precancerous lung lesion by regulating the IL-1β/miR-101/Lin28B signaling pathway." *Integrative Cancer Therapies* 16, no. 4 (2017a): 505–515.

Wang, X., Liu, P., Wang, F., Fu, B., He, F., and Zhao, M. Influence of altitudinal and latitudinal variation on the composition and antioxidant activity of polyphenols in *Nicoticana tabacum*. L Leaf. *Emirates Journal of Food and Agriculture* 29 (2017b): 359–366.

Wang, Yang, Chunmei Li, Liang Huang, Li Liu, Yunlong Guo, Li Ma, and Shuying Liu. "Rapid identification of traditional Chinese herbal medicine by direct analysis in real time (DART) mass spectrometry." *Analytica Chimica Acta* 845 (2014): 70–76.

Webster, Grady L. "Studies of the Euphorbiaceae, Phyllanthoideae II. The American species of Phyllanthus described by Linnaeus." *Journal of the Arnold Arboretum* 37, no. 1 (1956): 1–14.

Webster, Grady L. "Synopsis of the genera and suprageneric taxa of Euphorbiaceae." *Annals of the Missouri Botanical Garden* 81 (1994): 33–144.

Wei, Wan-Xing, Yuan-Jiang Pan, Hong Zhang, and Teng-You Wei. "Two new compounds from Phyllanthus niruri." *Chemistry of Natural Compounds* 40, no. 5 (2004): 460–464.

Wei, Wanxing, Xiangrong Li, Kuiwu Wang, Zuowen Zheng, and Min Zhou. "Lignans with Anti-Hepatitis B Virus Activities from Phyllanthus niruri L." *Phytotherapy Research* 26, no. 7 (2012): 964–968.

Yang, Baoru, Maaria Kortesniemi, Pengzhan Liu, Maarit Karonen, and Juha-Pekka Salminen. "Analysis of hydrolyzable tannins and other phenolic compounds in emblic leafflower (Phyllanthus emblica L.) fruits by high performance liquid chromatography–electrospray ionization mass spectrometry." *Journal of Agricultural and Food Chemistry* 60, no. 35 (2012): 8672–8683.

Yang, X., Liang, R.-J., Hong, A.-H., Wang, Y.-F., and Cen, Y.-Z. "Chemical constituents in barks of wild Phyllanthus emblica." *Chinese Traditional and Herbal Drugs* 45, no. 2 (2014): 170–174.

Zhang, J. T., B. Xu, and M. Li. "Relationships between the bioactive compound content and environmental variables in Glycyrrhiza uralensis populations in different habitats of North China." *Phyton-Revista Internacional de Botanica Experimental* (2011): 161–166.

Zhang, Lan-Zhen, Wen-Hua Zhao, Ya-Jian Guo, Guang-Zhong Tu, Shu Lin, and Lin-Guang Xin. "Studies on chemical constituents in fruits of Tibetan medicine Phyllanthus emblica." *Zhongguo Zhong yao za zhi= Zhongguo zhongyao zazhi= China Journal of Chinese materia medica* 28, no. 10 (2003): 940–943.

Zhang, Ying-Jun, Takashi Tanaka, Chong-Ren Yang, and Isao Kouno. "New phenolic constituents from the fruit juice of Phyllanthus emblica." *Chemical and Pharmaceutical Bulletin* 49, no. 5 (2001a): 537–540.

Zhang, Ying-Jun, Takashi Tanaka, Yoko Iwamoto, Chong-Ren Yang, and Isao Kouno. "Phyllaemblic acid, a novel highly oxygenated norbisabolane from the roots of Phyllanthus emblica." *Tetrahedron Letters* 41, no. 11 (2000): 1781–1784.

Zhang, Ying-Jun, Tomomi Abe, Takashi Tanaka, Chong-Ren Yang, and Isao Kouno. "Phyllanemblinins A– F, New Ellagitannins from Phyllanthus e mblica." *Journal of Natural Products* 64, no. 12 (2001b): 1527–1532.

Zhang, Ying-Jun, Tsuneatsu Nagao, Takashi Tanaka, Chong-Ren Yang, Hikaru Okabe, and Isao Kouno. "Antiproliferative activity of the main constituents from Phyllanthus emblica." *Biological and Pharmaceutical Bulletin* 27, no. 2 (2004): 251–255.

Zhao, Tiejun, Qiang Sun, Maud Marques, and Michael Witcher. "Anticancer properties of Phyllanthus emblica (Indian gooseberry)." *Oxidative Medicine and Cellular Longevity* 2015, (2015). Article ID: 950890, 7 pages, https://doi.org/10.1155/2015/950890.

Zubair, M. F., O. Atolani, S. O. Ibrahim, O. O. Adebisi, A. A. Hamid, and R. A. Sowunmi. "Chemical constituents and antimicrobial properties of Phyllanthus amarus (Schum & Thonn)." *Bayero Journal of Pure and Applied Sciences* 10, no. 1 (2017): 238–246.

Index

I

J

K

L

M

N

O

P

Q

R

S

T

U

V

W

Y

Z

Printed in the United States
by Baker & Taylor Publisher Services

Printed in the United States
by Baker & Taylor Publisher Services